ISBN 978-0-282-90395-4
PIBN 10872700

English
Français
Deutsche
Italiano
Español
Português

www.forgottenbooks.com

Mythology Photography **Fiction**
Fishing Christianity **Art** Cooking
Essays Buddhism Freemasonry
Medicine **Biology** Music **Ancient
Egypt** Evolution Carpentry Physics
Dance Geology **Mathematics** Fitness
Shakespeare **Folklore** Yoga Marketing
Confidence Immortality Biographies
Poetry **Psychology** Witchcraft
Electronics Chemistry History **Law**
Accounting **Philosophy** Anthropology
Alchemy Drama Quantum Mechanics
Atheism Sexual Health **Ancient History**
Entrepreneurship Languages Sport
Paleontology Needlework Islam
Metaphysics Investment Archaeology
Parenting Statistics Criminology
Motivational

7

8

9

10

11

FIG. 1.—Mammoth Hot Springs. Liberty Cap, Reservoir, and Garden.
FIG. 2.—Headquarters Building. FIG. 3.—Barn and Corral.

FIFTH ANNUAL REPORT

OF THE

SUPERINTENDENT

OF THE

YELLOWSTONE NATIONAL PARK.

BY

P. W. NORRIS,

SUPERINTENDENT.

CONDUCTED UNDER THE AUTHORITY OF THE SECRETARY
OF THE INTERIOR.

WASHINGTON:
GOVERNMENT PRINTING OFFICE.
1881.

TABLE OF CONTENTS.

LIST OF ILLUSTRATIONS.

FIFTH ANNUAL REPORT

OF THE

SUPERINTENDENT OF THE YELLOWSTONE NATIONAL PARK.

By P. W. NORRIS, SUPERINTENDENT.

OFFICE OF THE SUPERINTENDENT OF
THE YELLOWSTONE NATIONAL PARK,
Mammoth Hot Springs, December 1, 1881.

SIR: I herewith submit my report of operations for the protection and improvement of the Yellowstone National Park during the year 1881, with the request that, if approved, it may be printed.

Very respectfully,

P. W. NORRIS,
Superintendent of the Yellowstone National Park.

Hon. S. J. KIRKWOOD,
Secretary of the Interior, Washington, D. C.

SIR: As you are doubtless aware, the winter of 1880–'81 commenced very early, with unusual severity, and with attendant heavy snows throughout the United States, and continued so in those portions east of the Rocky Mountains. Such was not the case in the regions amid and beyond them, in which localities the latter part of the winter was very mild, followed by a continuously pleasant early spring. This condition of affairs resulted in a prematurely heavy rise in many western rivers, notably the Bighorn, Yellowstone, and Missouri, whose united waters, swept a resistless, devastating flood over a great agricultural valley, still robed in winter's mantle of snow. From these floods the elevation of the National Park preserved it, and allowed the mild but continuous daily rays of an unclouded sun to render the deep-sheltered glens and valleys luxuriant with herbage and fragrant with brilliant flowers, while the winter snows still rested low and chill, upon the mountain slopes above and around them. Rarely has man witnessed scenes more strangely mingling the weird and repellent with the charmingly beautiful, than these borders to fire-hole basins, or the sheen of the slanting sunbeams from ice-clad cliffs begirting geyser basins of spouting hot water, or the sulphur-lined fumerole escape-vents of smothered fires. These were among the scenes which greeted my return to the Park. In the East I had left the frugal farmer, with shelter, grain and care, nursing his starveling animals, hoping for the scanty herbage of a tardy spring; upon the Platte and the Great Plains I had seen the carcasses of thousands of animals claimed by the princely-improvident, fortune-trusting herdsman of the border, decaying where they starved, or frozen, fell; and, in the valleys of the Madison, the Gallatin, and the Upper Yellowstone, witnessed animals in only passable heart and flesh; while in the Park, at an elevation of 6,450 feet, in our little cliff-and-snow-girt valley, with its matchless hot springs, I found all our animals sleek, fat, and able, engaged in grading a road up the cañon of the east fork of the Gardiner River. In fact, the season in the valleys was as advanced by the middle of April, this year, as it was upon the 1st of July of last, and the roads in better condition; so that the limited amount of funds nuder my control available before the 1st of July alone prevented me from at once organizing my force and pushing improvements. Finding that, with the utmost prudence, I could keep only my assistant, C. M. Stephens, gamekeeper Harry Yount, and two additional men, I employed them in duties deemed most advantageous at this time. As Yount was no longer needed at the gamekeeper's cabin on the Soda Butte, and was a trusty person as well as an excellent hunter and scout, he was stationed at our headquarters (the gun-turret of which is a commanding lookout station) with instructions to daily scan, with a field-glass, our surroundings, visiting so much thereof as was convenient, paying attention to the repairing of roads and bridle-paths, and returning each night,. with game when needed, to the proper care of our buildings and other property. With the remaining men, one wagon, team, necessary tents and other outfit, I moved to our grade in the cañon of the East Gardiner, about four miles distant. By this plan we daily saved an eight-mile trip; in addition to which, I, being weary of city life, books and writing, the men of a winter's confinement to the house, and all longing for the freedom of camp life and blanket, these longings were thus gratified.

This cañon of the East Fork of the Gardiner is 3 miles long, 2,000 feet deep, with no egress from its vertical basaltic-capped cliffs, save by our bridges over the East and the Middle Forks, near their confluence, towards our headquarters, or past the falls and cascades to the Blacktail plateau. At this point we made our camp, moving it to little sheltered grassy nooks or glades, as we pushed forward our grade between the roaring torrent and the craggy cliffs. Here, beneath cloudless skies, the stately bighorn, the ferocious grizzly, and the royal eagle watched us from the cliffs, while grouse, deer, and elk were ever in sight and often within pistol-shot of our camp-fire; there were countless speckled trout in the dashing snow-fed stream beside it, and our quiet animals were half-hidden in pasturage and flowers. These scenes, with nights of refreshing sleep and days of cheering progress upon our new route from the cañons of the Gardiner—in short, our sports, our labors, and surroundings, all combined to render these bright camp-fire days among the most pleasant of those which I have ever spent in the mountains.

Late in May, with Harry Yount, I visited the Fire Hole regions, and besides noting geyser eruptions and removing fallen timber from the roads, planned and marked out much of our season's work in that direction, and thence alone proceeded to our gamekeeper's cabin on the Soda Butte. Meeting Rowland there, we together explored and marked a bridle-path to where checked by snow-fields upon the slopes of the Hoodoo Mountain; and on the route of return made interesting discoveries of Indian fortifications and fossil forests. A succession of long hot days early in June were telling rapidly upon the elevated snow-fields of the Sierra Shoshone range; each little rill joining its fellow, that another, thus deeply furrowing and undermining the softening ice-field or tottering crag, until launched a resistless snow-slide or avalanche of ice, rock, and crushed and tangled timber, with roar of thunder, crashing into the streams, rendering them for a time as turbid as the Missouri of the plains, and from their velocity and floating timber therein, far more dangerous to cross. I apprehend that three days of such experience as mine in crossing these mountain torrents while returning from the Soda Butte, would convey a more adequate conception of the resistless power of the mountain floods and their all-eroding effect upon the contour of mountains during countless ages, than the perusal of any work which has ever been written upon the subject.

Leaving the men to complete the cañon grade, I followed, noted, and sketched traces of the Indians, and of some unknown earlier occupants of these regions, from the Sheepeater Cliffs and Sepulcher Mountain, above the Mammoth Hot Springs, to the borders of the Park in the Yellowstone Valley. Then, while en route to Bozeman, for the purpose of selecting and forwarding our season's supplies, I improved my first leisure from urgent duty since 1875, in tracing and sketching such remains through the second cañon and Gate of the Mountains, upon the Yellowstone, together with the first terrace below, and the Bottler Park between the first and second cañons, a distance of fully 60 miles; and thence proceeded along the Trail Pass to Bozeman. These explorations are a continuation of those reported in my communication of 1877, as may be found under the head of "Prehistoric Remains in Montana," pages 327 and 328 of the Smithsonian Report of 1879. As it is my purpose (should there be no official objection) to furnish a fully illustrated report of these, and other traces, tools, weapons, and utensils of a supposed prehistoric people, to the Bureau of Ethnology, Smithsonian Institution, under the charge of Maj. J. W. Powell, I will here only state that they are deemed of peculiar historic interest and value. For the purpose of economizing valuable time, the latter part of June was devoted to hiring men, and the selection and forwarding by government teams of our supplies of tools, provisions and other outfit, for use after the 1st of July, and to be paid for with funds then available. From experience, I considered it best to work simultaneously upon both the Yellowstone and the Fire Hole routes from the headquarters, keeping up our communication there by weekly couriers. The men, wagons, teams, tools, and provisions were divided, and in readiness for an early start upon the morning of the 1st day of July, when, as before stated, our season's funds for improvement of the Park first became available. At dawn of that day the members of each detail were in harmonious but spirited rivalry for the start, anterior to which I read the names and duties of the various members of each, as well as the following address and instructions, furnishing a copy to the foreman of each party:

OFFICE OF THE SUPERINTENDENT, MAMMOTH HOT SPRINGS,
Yellowstone National Park, July 1, 1881.

MOUNTAIN COMRADES: Organized as we are for the protection and improvement of the Park, every member is expected to faithfully obey all the recently published rules and regulations for its management, and to vigilantly assist in enforcing their observance by all persons visiting it.

While labor in the construction of roads and bridle-paths will be our main object, still, with trifling care and effort, much valuable knowledge may be obtained of the regions visited, especially by the hunters and scouts, all of which, including the discovery of mountain passes, geysers, and other hot springs, falls, and fossil forests, are to be promptly reported to the leader of each party.

As all civilized nations are now actively pushing explorations and researches for evidences of prehistoric peoples, careful scrutiny is required of all material handled in excavations; and all arrow, spear, or lance heads, stone axes and knives, or other weapons, utensils or ornaments; in short, all such objects of interest are to be regularly retained and turned over daily to the officer in charge of each party, for transmittal to the National Museum in Washington.

P. W. NORRIS,
Superintendent of Yellowstone National Park.

FIRST PARTY, YELLOWSTONE ROUTE.

P. W. NORRIS, *in charge.*

Thomas Scott, foreman and wagon-master; George H. Phelps, hunter and scout; Julius Beltezar, packmaster; Clement Ward, cook; N. D. Johnson, Andy Johnson, Patrick Kennedy, R. E. Cutler, and Philip Lynch.

Supplied by one government wagon, four-mule team, and pack-train, the saddle animal of each man, and a good outfit of tents, tools, and provisions.

SECOND PARTY.—FIRE HOLE ROUTE.

C. M. STEPHENS, *Assistant, in charge.*

James E. Ingersoll, foreman and wagon-master; Harry Yount, game-keeper, hunter and scout; John W. Davis, packmaster; George W. Graham, blacksmith; Robert Clayton, cook; George Rowland, Frank Roy, Andrew Hanson, James Jessen, John Cunningham, Henry Klamer, Samuel S. Mather, Thomas H. Smith, George R. Dow, William Jump.

Supplied by one heavy wagon and a four-horse team, hired for the season. One medium-sized government wagon and two-horse team, with the blacksmith's forge, tools, and also pack-train, tents, and supplies; besides, as usual, each man with his own saddle animal, outfit, and weapons.

C. H. Wyman, my comrade of 1875, in the Soda Butte region, was left sole occupant of our headquarters save when George Arnhold, as for the past three years, made his weekly visits with the mails and supplies of articles as needed, and our couriers, who then received them for each party, and kept up a regular communication between them. As nearly all these men had shared the toils, privations, and dangers of the snowy pass, the weary watch, and the welcome camp-fire, and had been employed for their known worth and fidelity, either continuously, or during each season of labor, for from one to four or five, and one of them for eighteen years, they were truly comrades, treated and trusted as such, and are believed to be worthy of the above record of their names and respective duties. Although thus organized upon this occasion, such is now their knowledge of the routes which we have traversed, of each other, and of the various duties, that, aside from the assistant, blacksmith, and wagon-master, they could be reorganized in nearly any desired manner (and in fact were during the season with some addition to their numbers), without seriously impairing their efficiency; and I confidently challenge the mountain regions to furnish an equal number of men who, in the situation, circumstances, and peculiar difficulties under which we have labored, ever have shown, or are capable of showing, a better record of caring for public property or of making public improvements than is theirs.

Our day of starting being upon Friday, that day and the next, July 1 and 2, were spent by Stephens and party in repairing the grades and bridges to and beyond Willow Park, where they camped, spending the Sabbath and the national anniversary of the fourth in welcome rest and successful hunting; and as there were no intoxicating stimulants in camp, there was neither wrangling then, nor head nor heart aches when, with an ardent spirit of emulation in the performance of duty, they commenced the labors of the next morning. Important repairs and improvements were rapidly made at Obsidian Cliffs and the Lake of the Woods, and again repaired after a terrific water-spout (here called cloud-bursts), as well as at the Norris Geyser Basin and Gibbon Meadows; and the beautiful cone of a pulsating geyser, and some scalloped borders to adjacent pools, was, with much labor and difficulty, got out of a secluded defile two miles above the Paint Pots, for conveyance to the National Museum in Washington. Improvements were also made at Cañon Creek and other localities to and throughout the Fire Hole Basin. Thence, Stephens with his pack train reopened the great bridle-path, via the Shoshone and Yellowstone Lakes, to the Natural Bridge and Great Falls of the Yellowstone, returning by way of Mary's Lake to his wagons, and commenced pushing a road up the East Fork of the Fire Hole River toward the Great Falls. Meanwhile, I had with my party built a bridge over the East Fork of the Gardiner, at the head of its middle falls, another at the forks of the Blacktail Creek, there camping, with no other stimulants than the excitement of the use of rod and gun in securing a good supply of trout and elk meat, during the Sabbath and Independence Day.

We had ascended fully 2,000 feet by the only route possible for a wagon-road from the cañon of the Gardiner to the open, beautiful plateau of the Blacktail, whence a greater and more abrupt descent was requisite to reach the Yellowstone River, where Baronette's Bridge spans it from its East Fork to Pleasant Valley, this being the only place of approach through its terrible cañons, from 2,000 to 3,000 feet deep, between the Great Falls and the confluence of the Gardiner River, a distance of more than 40 miles. Previous long and careful research having failed to reveal a satisfactory route for a road, the 3d and 4th days of July were spent by Baronette, builder of the first house within the Park and the first bridge upon the Yellowstone River, and myself in a terribly trying but fruitless and final effort for a roadway through the yawning fissure region. Adopting a route which I had previously explored through an open pass in the Blacktail divide, we constructed a road with only a moderate amount of grade and bridging in passing between the vertical basaltic walls of a very modern lava overflow, and an impassable fissure-vent fully 1,000 feet deep, to Elk Creek, and through a geode basin to the famous "Devil's Cut," or Dry Cañon (as I more politely if less appropriately call it), to the stream skirting Pleasant Valley. While grading down the terribly broken banks of this stream we unfortunately broke our plow beyond repair by any person nearer than our blacksmith with Stephens, to whom our energetic wagonmaster, Scott, with a four-mule team and heavy wagon, took our broken plow and the fragments of another from our shop at headquarters to Stephens at Cañon Creek, exchanged it for the one with his party and returned, making the round trip of 100 miles within four days.

We were compelled to scale a sharp hill to escape an impassable cañon in reaching Pleasant Valley, and to traverse a boggy cañon to avoid a craggy cliff in leaving it, near the forks of the Yellowstone, and by steep grading and climbing reached the cliffs overlooking Tower Falls. Without sufficient time or means to construct a road into the yawning cañon of Tower Creek, we left our wagon and carried our plow into and across it above the falls; then attaching a span of mules, we plowed a furrow for a present bridle-path and one track of a proposed wagon-road over the lovely terraces, the grassy glades, and up the long foot-hill slopes of Mount Washburn to the snowy line within a mile of Rowland's Pass, which, in distance and elevation, is about midway between the foaming river, in the yawning cañon, and the storm-swept summit of the mountain crest. From this place Scott returned with the team, wagon, and two men to the Mammoth Hot Springs, where he quickly repaired the fences, filled our barn, besides securing a rick of excellent hay. They then hoed and irrigated our garden, and, with a supply of potatoes and other delicious vegetables therefrom, and supplies from Bozeman, proceeded to join Stephens in the Fire Hole regions. In the mean time, with the pack-train and the remainder of my party, I proceeded to greatly improve the bridle-path through Rowland's Pass, opened a new one two miles through timber, crags, and snowfields, to the summit of Mount Washburn, and, leaving the party to repair the bridle-path down the mountain and along the Grand Cañon to the Great Falls, I made a visit to Stephens and party, near the forks of the Fire Holes. Finding them energetically pushing the construction of a road towards Mary's Lake, I returned to my party, making ladders and various other improvements at and near the Great Falls, including a good bridle-path 5 miles below the falls to the roaring Yellowstone River in the Grand Cañon, where it is nearly 2,000 feet deep; and after planning and marking out a line of road, skirting Sulphur Mountain and the Mud Volcano, to the foot of Yellowstone Lake, united my party with that of Stephens.

After failing in a long-continued exploration for the discovery of a practicable pass through the Madison divide, towards the Yellowstone, we engineered a line of grade along its nearly vertical face, where little less than 1,000 feet high, and then through the cañon and along the route of General Howard in the Nez Percé campaign of 1877, to Mary's Lake. During the progress of this work, I embraced the first leisure of the season to visit the party of Justice Strong, Senators Sherman and Harrison, Governor Potts, the artist Bierstadt, and other gentlemen of prominence, accompanying them through the Fire Hole Basins, and with some of them—including Lieutenant Swigert, of Fort Ellis, in charge of their escort—to the Great Falls.

Prominent among the parties of visitors who were swarming to the Park early in August was that of Governor Hoyt and Col. J. W. Mason, the civil and military officers of regions embracing the Park, who were united in an expedition in search of a practicable pass for a wagon road from the inhabited portions of Wyoming to the National Park, of which they have a full appreciation and a pardonable pride. Having failed in a determined effort for the discovery of a pass at the head of the North Fork of the Wind River, after nearly a month of dauntless mountain climbing, they had just arrived at our camp, guided from the Two Ocean Pass by Harry Yount, whom I had sent to meet them.

Having been informed from Washington that want of funds would prevent the United States Geological Survey from making an exploration of these regions during this season, and deeming it very desirable to learn all possible regarding them in time for important legislation next winter concerning the Park and its boundaries, I accompanied Governor Hoyt, Colonel Mason, and party through the Sierra Shoshone Mountains to the head of the Great Cañon of the Stinkingwater, which they descended, while I completed the exploration, making important discoveries, and returned over the Soda Butte and Baronette's bridge from fearful snow-storms in the Goblin land, as will be shown under the head of explorations. While personally thus employed, and making but one brief visit to my men at the Mud Volcano, they, with highly commendable energy, completed a good road upon the line which I had laid out to the Yellowstone River, with a branch ascending it past the Mud Volcano to the foot of the lake, and another around the Sulphur Mountain to the mouth of Alum Creek, 4 miles above the Great Falls. They then returned through severe snow-storms, bringing in teams and outfit in good order to headquarters; and judging the employment of so large a force in autumn storms injudicious, most of the men were discharged, but provisionally engaged for next season if desired.

The remaining field operations with small parties were as follows: One with Davis, securing a fine collection of natural objects of interest and Indian relics from the fossil forests of the Soda Butte and Amethyst Mountain regions. Another was made, through severe snow-storms, to check vandalism and note geyser irruptions in the Fire Hole region, which was completed by Wyman and Rowland; another, by Stephens and Miller, in planning bridge sites and grades for next season upon the East Fork of the Yellowstone. My faithful gamekeeper, Harry Yount, having made his final tour and report, tendered his resignation; records of all of which will be found in their proper order.

The final trip of the season was made with the teams in October, hauling out to Fort Ellis, Montana, a large and valuable collection of natural and anthropological objects of interest for the National Museum in Washington; and then to Bozeman, 4 miles distant, for the purpose of closing the business affairs of the season, and the purchasing and forwarding of our winter's supplies.

Our buildings are well repaired, and wagons, tools, and other outfit secured for winter; during which it is my purpose to retain only my trusted assistant, Stephens, and Packmaster Davis, for the care and protection of the building, animals, and other public property.

The season for profitable labor in the Park closed, as it had commenced, unusually early; but the practical knowledge which has been acquired of the climate and peculiarities of these regions, the careful protection of teams, tools, and provisions, the excellent character and organization of my men, enabled me to make large and substantial improvements, and win the approbation of the candid, practical portion of the numerous and prominent tourists to the wonder-land. Neither myself nor others are as well satisfied with the season's protection of the forests from fire, or the geyser cones or other objects of natural interest from vandalism; all of which, with my suggestions as to a practical remedy, will be found in appropriate sections of this report.

The unavoidable failure of all my aneroid barometers to register correctly is a source of deep regret and a serious loss; but the thermometer readings, which have been regularly and carefully noted and preserved at the Mammoth Hot Springs during the entire season, as well as during my explorations of the Rocky and Sierra Shoshone Mountains, and those of Wyman in the Geyser basins, it is believed will be perused with interest, as greatly increasing our meager knowledge of the peculiar climate of those regions.

AREA OF THE PARK.

Two matters in connection with the Yellowstone National Park tend to great and general misapprehension regarding it. These are, first, its name, and second its area; or, as are perhaps best treated, inversely.

The large, beautiful, and (so far as then explored) correct map by Henry Gannet, M. E., topographer of the United States Geological Survey of the said Park during 1878, now in press, shows it to be an oblong square, 62 miles in length from north to south and 54 miles in width from east to west, containing 3,348 square miles. The extra census bulletin, by Mr. Gannet, now geographer of the tenth census of the United States, under date of September 30, 1881, page 4, shows that the area of the State of Delaware is 1,960 square miles; State of Rhode Island, 1,085 square miles; District of Columbia, 60 square miles; and page 17 of said bulletin shows the aggregate area of the counties of New York, King's, and Richmond, of the State of New York, is 150, equal to 3,255 square miles. Thus the most recent and reliable authorities extant show that this great national land of wonders contains 93 square miles in excess of the aggregate area of two of the original thirteen States of the Union, the District of Columbia, containing the capital, and the three counties of the State of New York, which embrace the commercial emporium of the first and third cities of the nation, having an aggregate population of about 2,500,000. Nor is this a full statement of the case; as, if to this account were added the actual excess of surface measurements of this peculiarly broken region, over those relatively level eastern ones, it would (see bulletin, page 4) certainly exceed that of Connecticut, 4,845 miles, and, with the adjacent Goblin Land and other regions which I have explored during the past two seasons, fully equal that of New Jersey (bulletin, page 4)—7,455, or Massachusetts (same page)—8,040 square miles, or several other of the original States of the Union.

Prominent among the bordering points of observation of this vast region is Electric Peak, near the northwestern border, elevation 11,775 feet; Mount Norris in the northeast, 10,019; Mounts Chittenden, Hoyt, Langford, Stephenson, and others in the eastern Sierra Shoshone border, and Mounts Holmes and Bell's Peak upon the western, ranging between 10,000 and 11,000 feet high, and Mount Sheridan, near the southern border, 10,385 feet high, still backed by the Grand Teton, landmark of all those mountain regions, which is over 13,000 feet in height. But Mount Washburn, towering upon the brink of the yawning Grand Cañon waterway of the Yellowstone Falls and Lake, 10,340 feet high, is the most central, accessible, and commanding for a general view of the park and its surroundings. From its isolated summit can be plainly seen on a fair day, as upon an open map, not only this lake and cañon but many others also; countless flowery parks and valleys, misty sulphur and steaming geyser basins, dark pine and fir-clad slopes, broken foot-hills, craggy cliffs, and snowy summits of the sundering and surrounding mountains. No tourist should fail in securing this enchanting view, the best plan of obtaining which is, upon reaching the meandering rivulet-fed lawns of the Cascade, the Glade or the Antelope Creeks, to go into camp, and await the dawn of a cloudless summer's morning. Then, to the scientist, the artist, or the poet, and to the weary and worn pilgrims of health and pleasure, from our own and other lands, ardent to secure the acme of mountain-climbing enjoyment, or in viewing the lovely parks and yawning cañons, the crests of glistening ice and vales of blistering brimstone, the records of fire and flood, the evidences of marvelous eruptions and erosions of the present and the past, and day-dreams of the future in the commingling purgatory and paradise of the peerless Wonder Land of earth, I would say, leisurely ascend the terraced slopes of Mount Washburn, and from its oval summit, with throbbing heart but fearless eye and soul expanding, look around you. One day thus spent would more adequately impress the mind with the magnitude and marvels of the Park, and the vast amount of exploration and research necessary in finding routes, and the enormous amount of labor and hardship unavoidable in the construction of buildings, roads, bridle-paths, trails, and other improvements, even when unmolested by hostile Indians—as during the past two years only—than a perusal of all the reports and maps of the Park which have ever been published.

Owing to the lack of natural curiosities worth retaining, in the three-mile strip of the Crow reservation in Montana, upon the north, or the four-mile strip in Montana and Idaho upon the west, the desirability of having the entire Park under one jurisdiction, as well as for other and weighty reasons fully set forth in my report of 1880, I again earnestly recommend re-ceding to the jurisdiction of those Territories all of the Park not embraced by the now surveyed northern and western bound-aries of Wyoming, leaving to future explorations and development the fixing and surveying of the remaining borders. It is hoped this may be done next season by the United States Geological Survey.

This necessarily lengthy explanation of the first question as to the mag-nitude of the Park so nearly disposes of the second, as to the name, that I only add that although it is so vast and broken by mountains and cañons into countless partially or wholly isolated parks and valleys, still the whole of it is nearly encircled by snowy mountains with few passes, being thus park-like in character, and the name correct, or at least diffi-cult to substitute by one more appropriate.

THE TWO MAIN APPROACHES TO THE PARK.

The explorations of myself and others, previous to my assuming the superintendency of the National Park, led to the correct conclusion that there were only two natural valley routes of access for wagon or rail-roads thereto, viz: the one up the Yellowstone River to the initial point on the northern boundaries of the Park, at the confluence of the Yellow-stone and the Gardiner Rivers, some five miles below the Mammoth Hot Springs; and the other from the West via Henry's Lake and the Upper Madison River to its head at the confluence of the Fire Hole Rivers. The elevated passes over the Rocky and Sierra Shoshone ranges will be noted in their proper connections.

EASTERN APPROACHES TO THE PARK—THE VALLEY OF THE UPPER YELLOWSTONE AND THE TWO OCEAN PASS.

There are many and important indications that the towering lava cliffs which border the Yellowstone Valley above the lake were once lashed by the waves of its then extended little finger, fed by mountain torrents in yawning gulches, and drained through Two Ocean Pass into Snake River and to the Pacific Ocean, much as the ancient lake Bonneville (of which Salt Lake is a dwindled remnant) once drained through the Porte Neuf Cañon; and that the present Yellowstone and Bridger's Lakes, as well as the deep blue alpine-like appearing waters of the Upper River between them, are only remnants of this matchless mountain lake, since a less elevated outlet was elsewhere worn. Two Ocean Pass is either a natural gap or a broadly and smoothly eroded pass directly through the continental divide, trending from Bridger's Lake, near the head of the ancient one, southwesterly towards Jackson's Lake, at the foot of the Grand Tetons. Some 4 miles from the main valley this becomes a smooth open marshy meadow, fully half a mile wide; for the first 6 miles of which the waters creep sluggishly towards the Yellowstone, and then, in like manner, towards the Snake River. From these circumstances, the first slope is called the Atlantic, and the last the Pacific Creek; and are both fed along their courses by torrents from the snowy mountains upon each side as usual, the only novel feature heretofore known of this, being that one of these streams from the south enters the pass so near the sum-mit that portions of its snow-fed waters discharge through these creeks towards both the Atlantic and Pacific Oceans, and hence the names of those creeks, the side creek, and the pass. Our camp of this year was made upon the left-hand side of the Pacific Creek, where a comparatively mod-ern overflow of lava has not only pushed encroaching basaltic walls far into the pass from the north, but a narrow stream, of the same material 20 or 30 feet in thickness, entirely across, and for a time severing it and forming the summit and divide of the pass. Through this, from erosion or other causes, two openings have been formed. I had never, from record or narrative, heard of a creek upon the north side, nor had I specially observed it until in crossing the mountain towards Barlow's Fork of Suake River I found that while the small but permanent and uniformly flowing Two Ocean Creek drained a snowy basin high above, but within a mile or two of the pass, a much larger one, in fact a fair-sized mill-stream, cuts a yawning gorge in descending over 2,000 feet within 4 miles from the snowy summit of the Rocky Mountains to the north of the pass. This enters directly opposite the other creek, a knowledge of which at once solves the whole mystery which has always shrouded this pass; for with but one feeder, no matter what its angle of entrance to the pass, it would have, as is commonly the case, cut and followed a channel to one ocean, not both, but, with both torrents cutting their gorges and depositing the débris directly opposite, a broad dam has slowly but steadily accumulated entirely across the pass (there less than a mile wide) from the convex or sloping ends and sides of which the streams, broken into smaller channels by the ever descending and changing masses of rock and timber, actually does divide the waters, and portions of each flow through thousands of miles of yawning cañons and mighty rivers to opposite oceans. Although, during this year, a somewhat larger portion of these waters drained into the Atlantic, there is a liability to fluctuation naturally, and little labor would be necessary each season to throw all of these waters, from off this sloping divide, into their former course to the Yellowstone, or through these two openings in the former lava divide, 200 yards upon the Pacific side of it.

In search, not of a better pass or approaches than that at the matchless "Two Ocean," but rather a shorter and better route than the one through dense, and, for the most part, fallen timber, through Trail Pass and by the fingers of the Yellowstone Lake, we scaled the main divide, and, shivering in the snow among the clouds, searched our maps and scanned the surroundings, especially those upon the desired route northwesterly. The scene was grand and inspiring, but the practical part of it was that we could distinctly trace the Grand Tetons, Mounts Sheridan, Hancock, and other familiar snowy peaks, with traces of the numerous fountain heads of the Snake River, and their valleys or cañons, and notably the main one, the Barlow Fork, apparently to our feet, and the desired pass in the main range to Pacific Creek, some miles below us. Buoyant with hope of a warmer region, we frightened scores of big-horn elk and grizzlies, in an impetuous descent of over 2,000 feet into a deep, narrow valley, connecting the Falls Fork of the Yellowstone with a stream which we were rapidly descending, hopeful of a nooning in the lovely Barlow Valley, when, with a sudden turn to the left, it cut directly through the mountain to the Pacific Creek, leaving us to follow the Barlow when we could find it. This we did by way of a pass and mountain spur, which certainly could not have been visited by Jones or Hayden, as neither these nor other portions of a region 6 or 8 miles in width are represented upon the maps of either of these gentlemen. But, as elsewhere stated, the pass to Fall Creek is evidently that traversed by Phelps in 1864, and hence given his name.

A thorough exploration of the region between the Barlow Valley, Mount Sheridan, and Heart Lake to Riddle Lake and the fingers and thumb of the Yellowstone, renders it evident that the route as proposed by Captain Jones and Professor Comstock, in 1873, and by Governor Hoyt and Colonel Mason this season, from Wind River over Tog-wa-tee Pass to the Buffalo Fork and Pacific Creek, waters of the Snake River, can utilize the old Two Ocean and Upper Yellowstone route, or a new one through the lower end of Phelps Pass, and a side one from it, through which we reached the magnificent timber and charming valleys of the Barlow and the Heart Rivers, and the low timbered plateau summit of the Continental divide where there is no mountain, past Lake Riddle, to connect with our bridle-path from the Fire Holes and Shoshone Lake at the western end of the Thumb of the Yellowstone.

I may here add as an objection to the adoption of a water-shed as a boundary of the park, that in this exploration between Phelps Pass and Heart Lake, I traversed the main continental divide, following a tolerably direct course, no less than eleven times in one day.

The interlocking fountain-heads of stream in the Sierra Shoshone range render its water-shed equally tortuous and objectional.

NEW PASS OF THE SIERRA SHOSHONE RANGE.

The narrow elevated pass discovered by Captain Jones in 1873, south of Mount Chittenden, several similar ones explored by myself at various times north of it, and Sylvan Lake, discovered, named, and sketched, together with its supposed drainage, as correctly as possible in a snowstorm, by members of the Hayden expedition of 1878, was all known of passes in the entire Sierra Shoshone range prior to this season. From mountains at a distance I had often observed a deep depression in the serried crest of this range which could not be seen when among its broken foot-hills. The length of time expended by Governor Hoyt and Colonel Mason in their outward route from Wind River would not allow of the search for a pass there, in our crossing to the Stinkingwater, or while following it to its great cañon, which they descended, leaving me to prosecute the exploration. This I did, ascending several creeks, and from lofty peaks viewing all the others, as well as passes of the range above the cañon, finding few trails and no practical passes until on the north bank of the second creek below Jones's I found an ancient but very heavy lodge-pole trail, which I traced eight miles to the forks of the creek, and camped in a grove of cottonwood and other timber—indicating a sheltered and warmer location than is common at that elevation—and some pine trees 150 feet in height. Phelps caught trout, Roy kept camp and cooked supper, while Yount ascended the south and I the north fork of the creek. He reported impassable, snowy barriers; myself, indications of a pass some 5 miles distant; and the evening with the glistening of a glorious sunset and the haloes of the harvest moon of other lands upon the Giants' Castle, towering athwart the glittering stars, was spent in plans, preparation, and hopes of a morrow's crossing of the divide.

Pressing ahead of the packs in the morning, I was blazing the trail along the steep acclivity, when it dwindled, and, in Shoshone guttural, *kaywut*; or, according to border provincialism, "played out," and a sharp turn to the right at once revealed the cause to be the branching of the trail for various elevations in ascending to a low, clear-cut, but very narrow pass directly through the range, unlike all others, which are elevated, with very steep, rocky climbing from one or both approaches to the sharp, narrow crest. We reached the summit in time for a romantic noon camp on a velvet lawn of grass and frost flowers, beside an Alpine lake supplied by a snow-fed rivulet, skipping in several fifty-feet leaps from the cliffs; and as meat was wanting, Yount killed a blacktail, myself an elk, the surplus of which, and want of other provisions, caused the return of Phelps and Roy, with the most of it and all the pack animals save one each for Yount and myself, to our main camp at the Mud Volcano, they not returning to us. This pass has more the appearance of a natural gap, not quite closed by two mountains of eruption, than by the erosion of a narrow pass; but whatever the cause, it is a very low, direct one, with good approaches for a trail or wagon road, the only drawback being sev-eral heavy mountain slides, some very ancient, and others of compara-tively recent occurrence, the latter with immense masses of angular rocks filling it for at least a mile from fifty to two hundred feet deep, and the former causing a chain of three lakes, the most western of which is evi-dently the Sylvan Lake of Hayden's map of the explorations of 1878. This is shown correctly, but not its drainage, which I did not find; but, as the next lake in the pass drains toward this, its outlet cannot be to the Stink-ingwater—as the one at the cascade probably is—but even this only by percolating through these modern rock-slides. As this pass is nearly abreast the eastern side of the Yellowstone Lake, affording a fine route *via* Clear Creek to and a route each way around it, and there appears to have been comparatively little recent rock-sliding in the pass, it seems to promise its old pre-eminence as such of the range, by the making of a rocky road, as I did at the Obsidian Cliffs in 1878, over that portion of the pass which doubtless caused its abandonment by the Indians for at least a generation. In reply to my pressing inquiry of We-saw regarding a pass in that direction, while upon the range going out, his only answer was a French-like shrugging of the shoulders and ejaculation, " *Me no go there ; maybe Bannock Indian, long time 'go.*"

DIRECT CONNECTING ROAD.

One of the early and important plans of the park was the exploration and opening of a line of wagon road, upon the most direct practicable route, from the headquarters across the park to and through the other entrance thereto, thus connecting them for the convenience of our laborers, the public, and the military for their protection.

Important explorations were made in 1877 upon my route of 1875, and were completed and a rough road opened during the Bannock raid of 1878. This was somewhat changed and shortened through the earth-quake region, in order to meet the new entrance over the Plateau of the Madison instead of through its cañon, in 1880, and with the improve-ments since made at Cañon Creek and elsewhere only requires important grades to save crossing the Gibbon in its cañon, and opening of the routes through the Middle Gardiner Cañon, to render it a direct and permanent route connecting the two main entrances.

CIRCUIT OF ROADS.

Another improvement contemplated in the first general plan of devel-oping the park, and which, though often delayed, has never been aban-doned nor forgotten, but persistently pushed at every opportunity each year, has been the construction of a bridle-path upon a route to be mainly followed by a wagon road connecting these two main entrances, from the Mammoth Hot Springs via the Forks, Great Falls, and Lake of the Yellowstone, to the Forks of the Fire Holes, so that tourists could ultimately enter the park by one of these main approaches, visit the principal points of interest with wagons; those of less importance by branching bridle-paths, leaving it by either. Bridle-paths were early opened, and important changes made, with exploration and oppor-tunity, until the whole line was planned, and although the greater part of 1880 was unavoidably devoted to opening the new route over the Madison Plateau instead of its cañon, still, a good start was made in the cañon of the East Gardiner River, from the Mammoth Hot Springs at one end, and up the East Fire Hole River from their forks at the other, during 1880; and the main improvements of this season have been in the construction of this line of road from both ends. As elsewhere stated, the remarkably favorable spring of this season would have permitted the advantageous use of a much larger appropriation than was at my command, but what I had was promptly and prudently expended in the warm sheltered cañon of the East Gardiner.

After July 1, when this year's appropriation became available, until the untimely heavy snows of September rendered such field-work injudicious, the construction of this road was pushed with a vigor, skill, and success, resulting from thorough previous exploration, preparation, and experience, aided by a reliable and active assistant and force of veteran laborers, well understanding their duties and emulous in surmounting the attendant difficulties of climate and surroundings.

The proposition of responsible parties to introduce a portable steam saw-mill for the purpose of sawing lumber for a steamboat upon the Yellowstone Lake, hotels at its foot, and falls of the river, as well as for the government in the construction of bridges, added to the necessity of reaching the foot of the lake this season. After the construction of bridges, culverts, and grades in the open valley of the East Fire Hole, much of which was boggy, and the failure of long and laborious exploration to reveal a practicable pass through the precipitous Madison Divide, it was crossed by a long and uniformly excellent grade along its nearly vertical face to the narrow, dry cañon outlet of the ancient Mary's Lake, along the grove-girt border of its clear but brackish waters, uninhabited by any kind of fish, through the adjacent noisome sulphur basin to the deep valley and grassy lawns of Alum Creek. Thence, winding amid the bald, eroded, and still eroding hills of a short divide, down the open meadows of Sage Creek to the old trail near the Yellowstone River, midway between Sulphur Mountain and the Mud Volcano. From there, one branch was pushed up the river past the Mud Volcano, Nez Percé Ford, and a succession of enchanting groves and flowery lawns, beside the broad, placid, blue waters of the peerless Yellowstone, to Toppin's Point and miniature harbor at the foot of its lake. The other branch was constructed by winding ways, amid verdant hills, passing the stifling fumes of Sulphur Mountain, to the mouth of Alum Creek, four miles up the Yellowstone, above its Great Falls. The other end of this circling line of road was forced through the cliff-walled cañon of the East Gardiner, the grassy plateaus and lava beds of the Blacktail, beside the yawning, impassable fissure vents fronting Hellroaring Creek, through the Devil's Cut (which I am trying to rechristen Dry Cañon), and down the mountain slopes fully 2,000 feet to Pleasant Valley and the Forks of the Yellowstone, in this only practicable gap of the Grand Cañon for a distance of more than 40 miles. By careful research, we carried our road to the summit of the cliffs overlooking alike one of the finest views of the Grand Cañon, the Tower Falls, and the meeting of the foaming blue waters between them. This leaves a gap of less than 20 miles in distance between the Tower Falls and the terminus of the other end of our road at the mouth of Alum Creek, and hence the completion of our much-desired circuitous line of road to the main points of interest in the Park, situate west of the Yellowstone Lake and its Grand Cañon. As before shown, the two main routes of access, as well as the direct or Norris Geyser Basin route, being open, this little gap is all re-

2 Y P

maining to complete the plan of roads originally adopted and persistently adhered to through vexatious difficulties, and delays, and annoying public misapprehensions.

Although this gap is so short and some portions of it an excellent natural roadway, yet the yawning cañon of Tower Creek, with its vast amount of rock-work, culvert, and bridging above the Falls, the scaling of Mount Washburn through Rowland's Pass, extensive bridging, timber-cutting, and grading along the Grand Cañon and near the Triple Great Falls, together with the absolute necessity of several small bridges and extensive grading, or twice bridging the Yellowstone above the Falls, to connect with the other road at Alum Creek, renders it incomparably the most expensive of any equal portion of the route, and hence it was left until the last; and $10,000 is deemed necessary, and is specifically recommended to be appropriated, for these purposes during the coming fiscal year. This sum, in addition to the amount annually appropriated, might perhaps complete this road, were all others neglected. But this would appear injudicious, as, although the road over the Madison Plateau is deemed an excellent one, save the grades at each end, and *they* as good as are possible to have been made there, with the limited time and means at my command when this was done, still, they are very steep for hauling heavy boilers or mill or steamboat machinery, and need extensive change of grade, or else of the entire line, and returning to the circuitous Cañon route, with its unavoidably long and expensive grades, or bridging, or both, and which cannot properly be longer delayed. With nearly equal force, this necessity pertains to the extension of the road up the East Fork of the Yellowstone and Soda Butte, as the only route to the gamekeeper's cabin, the fossil forests, medicinal springs, and extension to the borders of the park, of a very important at least bridle-path route via the Clark's Fork mines to the Big Horn Valley and Fort Custer.

There is also a necessity for important bridle-paths up the East Fork Valley to the Goblin Land, and by a newly-discovered pass to Pelican Creek and Steamboat Point, on the Yellowstone Lake. This route also necessitates the purchase of the Baronette Bridge, recognition of it as a toll bridge, or building another, with better approaches, near it. The great desirability of constructing a road via the Middle Gardiner Cañon is believed to be rendered evident in the section devoted to that subject. Nor should the views of the governor, the military officers, and leading citizens of Wyoming Territory, in which the park is mainly situated, their explorations for a route to this Wonder Land, and their efforts to open it, as elsewhere explained, be ignored, but at least a substantial bridle-path route should be opened from some of ours to the borders of the park near the Two Ocean Pass, or via the new one which I explored during the past season through the Sierra Shoshone Range to the Great Cañon of the Passamaria, or both of them. In this connection I may state that my former knowledge and this season's explorations alike sustain the views of Governor Hoyt and Colonel Mason as to the practicability and necessity of a wagon-road from the Wind River and Two Ocean or the Stinkingwater (Passamaria) route to the park; and, as such, I do most cordially indorse their report favoring the appropriation of a sum sufficient to open a good wagon-road from the Wind River Valley or from the Stinkingwater to the borders of the park.

CAÑON OF THE GARDINER RIVER.

In addition to long, yawning, and interesting cañons upon all of the forks of the Gardiner River, high in the snowy ranges not traversed by any of our roads or trails and hence not necessarily mentioned here, there are four of great interest and importance within five miles distance and in plain view of our headquarters at the Mammoth Hot Springs, viz: One upon each of the three forks, or branches, cut in their precipitous descent of nearly 2,000 feet down the basaltic cliffs to our deep sheltered valley, by them eroded in some remote period, and another carved fully 1,000 feet deep by their united waters in escaping to the Yellowstone. Winding along the western terraces above the latter cañon, we have constructed our road to the main Yellowstone Valley, also one over the elevated Terrace Pass, around that portion of the cañon of the West Gardiner—which is utterly impassable for even a game trail—on our road towards the Fire Holes and through the beautiful cañon of the East Gardiner, ornamented by basaltic column-capped cliffs above and around the falls and cascades, on our road of this season to the Forks of the Yellowstone. The remaining cañon of the middle, and far the largest, fork is utterly impassable, but a bridle-path was made in 1879 along the precipitous face of Bunsen's Peak above it as preliminary to a road line. This bridle-path, as stated in some preceding report, has been in practical use and has demonstrated the feasibility of the route for a road to connect with that to the Fire Holes near Swan Lake. With no increase of distance this route will save several hundred feet in elevation, afford a picturesque view of the Mammoth Hot Springs, government buildings, and sheltered cliff-girt valley from one end of the pass, the upper valley with its rim of snow-capped mountains from the other, and within it the Sheepeater Glen, the vertical walls and uniquely interesting rotatory or fan-shaped basaltic columns, the roaring falls and splashing cascades of the Middle Gardiner, in wild, majestic beauty second only to those of the Grand Cañon of the Yellowstone, in the Wonder Land. Long and careful search and engineering resulted in the selection of a route along our timber road to a terrace overlooking the lower cascades of the West Fork of the Gardiner, which is to be crossed upon a short but very high timber bridge, and thence by a moderate and uniform grade along the pine clad face of Bunsen's Peak to the summit of the pass, amid the spray and thunder of a cataract nearly 200 feet high, in an eroded cañon more than 1,000 feet deep—a route combining so much of surpassing interest and practical value that only the want of means to divert from the pressing necessity of opening new routes to the Great Falls and other leading points of attraction has prevented its construction, and will insure it, with the first means at my command to properly thus expend.

MOUNT WASHBURN BRIDLE-PATH.

Successive seasons of exploration and research have resulted in the partial abandonment of the old route, with its several steep ascents upon the cold snowy side of Mount Washburn, the gulches of Dunraven's Peak, and the beautiful, but, in places, boggy valley of Cascade Creek, for the bridle-path route of a road ascending by long, easy grades from the pleasant meadows of Antelope Creek to the elevated but only summit of the route, in Rowland's Pass, and thence in like manner down its warm sheltered face to the grassy glades and sulphur basin, between it and the Grand Cañon, and skirting the latter, with its matchless scenery, to the Great Falls. An easily accessible peak upon the very brink of the Grand Cañon, about half a mile east of Rowland's Pass, affords a commanding view of it in all its windings and yawning side cañons, from the Forks to the Great Falls of the Yellowstone, and the terribly eroded, gashed, and repellant-looking unexplored region beyond it. By a short moderate ascent west from the summit of the pass, an open spur is reached, which, in less than two miles of gradual ascent, scales the highest peak of Mount Washburn if desired, although it is but little more elevated and commanding than portions of the snowy crest before reaching it.

PAINTED CLIFFS—BRIDLE PATH INTO THE GRAND CAÑON.

This path leaves the main one, from Mount Washburn, at the eastern end of an open marsh, about 5 miles below the Great Falls, and, passing fully a mile through an open pine forest, reaches the head of the cañon, and winds along the face of a mountain slide to the small, but beautiful and noisy, Safety Valve pulsating geyser, situated in the narrow valley between this slide and the mountain face. For a proper understanding of this location it is necessary to explain that, evidently at a comparatively recent period, the eroding river and the erupting fire-holes along it have undermined portions of the nearly vertical walls, some of which are fully a mile along it and nearly half as wide and high, precipitating them into and damming it until cut asunder by the resistless current of the foaming river, often leaving long portions of these enormous mountain-slides with the timber undisturbed upon them. It thus presents the appearance of a lower bank, or terrace, with a nearly vertical face of the peculiar ancient lake formations of this region, above and below it. Along the line of contact above this mountain slide, skirting the river below, and at the terribly ragged ends of it, is a line of noisy escape vents of smothered fire, of which one is the "Safety Valve," thus named at its discovery last year, from its powerful and distinct reverberations along the cliff, which were then much more audible than during this season. This is nearly a mile in distance, and 1,000 feet in descent, below the summit of the cliffs, or one half of the entire distance and descent in the cañon, the lower half of which was made through a line of mingled active and extinct and crumbling geyser and other hot-spring formations, along the ragged edge of the lower end of the mountain slide to the foaming river drainage of the mountain snows. This stream we found literally filled with delicious trout of rare size and beauty, and so gamy that all desired of them were caught at each of our visits of this year, during our brief nooning, using as bait some of the countless salmon flies which were crawling upon the rocks or on our clothing, upon hooks fastened to one end of a line, the other being merely held in the hand or attached to some chance fragment of drift-wood; but the sport seemed harder upon the hooks and lines than upon the trout, which were abundant, both in the river and out of it, after the loss of all our lines. Although this is strictly true in our experience, it is but just to state that some other persons who were there at a later hour of the day or period of the season, while seeing countless trout, found them less voracious.

The beautiful tinting of the cliffs in this locality, not unlike beauty elsewhere, seems only skin-deep; i. e., the material beneath is often nearly white, and the brilliant coloring only brought out by surface oxidation of the various mineral constituents; and, although not deeming our path dangerous, I would suggest that anglers who may visit this place should not become so engaged with the beautiful speckled trout as to forget that their charming lady companions may need their nerve and assistance in the horseback ascent of the cliffs. Here, only, between Tower Creek and the Great Falls of the Yellowstone, does a

CRYSTAL FALLS AND GROTTO POOL, WITH BRIDGE AND LADDERS.

bridle-path reach the foaming, white surfaced, ultramarine blue waters of the "Mystic River," and the long, horizontal, cornice-like grooving of its clearly banded and rainbow-tinted walls and tottering cliffs; in short, the seclusion, the scenery, and the surroundings of this hidden glen of the Wonder Land render it one of the most uniquely attractive so that the few tourists who fail to visit it will never cease to regret their neglect.

THE TRIPLE OR GREAT FALLS OF THE YELLOWSTONE, AND THE BRIDLE-PATH AND TRAILS THERETO.

These, as is well known, are the Upper Falls, of 150 feet, or about the same height as those of Niagara; the Lower Falls, nearly one-half mile below, of about 350 feet; and upon the west side of the river, midway between them, the Crystal Falls, or Cascades of Cascade Creek, near its mouth, in height about equaling the Upper Falls. Upon the very brink of the latter the main bridle-path to the lake passes, affording a fine view of them—the foaming rapids above and the rippling river below them— to the head of the Lower Falls, which is reached by the 500-feet descent of a good trail from the main one, or bridle-path, which crosses the creek upon a good bridge constructed last year from two projecting trachite rocks, nearly 40 feet above the famous Grotto Pool, between the upper fall, of 21 feet, and the lower, of more than 50 feet, beside a leaping cascade below it. This pool is caused by the sheet of water in the upper fall being at right angles with the stream, thus facing and undermining the eastern wall, and beneath it forming a broad, deep pool of placid water, nearly hidden under the narrow shelf of rocks between the two leaps of the cataract, and from its peculiarities named by me, in 1875, Grotto Pool. From a pole railing to the cliff between the bridge and the brink of the cliff overlooking the lower leap I this year placed a substantial, well-supported ladder to a projection of the cliff, and from there another to the foot of the Grotto Pool, and also some benches, for the convenience of tourists, beneath an overhanging rock and the lofty bridge along the narrow way between the wall and the water beside it. (See Fig. 1.) A sudden but violent hail and thunder shower, peculiar to mountain regions, compelled us to utilize this newly-reached shelter before leaving it, and for a brief period the flood-gates of heaven and the torrents of earth, with their mingled thunders, combined in a carnival of surging elements and waters above, beneath, and beside us.

Near the rustic bridge spanning Spring Creek a long, rough, and dangerous trail descends to the foaming river, and a bridle path ascends to the Point Lookout Cliff, 1,000 feet above it, about one mile below and directly fronting the lower fall, inviting, within its barricade border of poles, the finest safe view of them from any quarter. From here the great notch near the northern and two smaller ones near the southern edge of the clear-cut and formerly smooth water-line of the fall are evident, at a glance, to any person familiar with the falls or photographs or correct sketches of them prior to this season. The detachment of great masses of the rocky face of the falls is the cause, but only the commencement of what will in time follow, and ultimately change the appearance of these falls, but probably not their aggregate height.

NATURAL BRIDGE AND BRIDLE-PATH TO IT.

Since the first description of this interesting freak of nature, which is situated near the Yellowstone Lake, was published on pages 22 and 23 of my report of 1880, I have so changed the route of the bridle-path as to invite an excellent view of the archway at several points of observation within the distance of a mile below or fronting it, and then, after cross-ing a warm creek near some beaver ponds, ascend by a winding way to and across it. Thence the trail, within a distance of two miles, descends through a beautiful pine forest, meanders along the shore of the nearly severed extension of Bridge Bay, and across some lovely grove-girt lawns to the old route upon the shore of the lake. The danger of a general conflagration alone deterred me from burning out several miles of nearly impassable fallen timber, thereby materially shortening the trail to the thumb of the lake. No other substantial natural bridges over a perma-nent water-course have been discovered, but several wind and storm worn tunnels, high amid the tottering crests of the Sierra Shoshone Range, were found and sketched; also one between the first and second peaks from the southwestern slopes of Mount Norris, nearly fronting the famous extinct geyser cone of Soda Butte, although high above and scarcely perceptible from it, but showing a clear cut outline of blue sky directly through the craggy crest, from the great terrace of Cache Creek. At that distance, and even nearer, this opening so closely re-sembles the adjacent snow-drifts that Rowland, who was with me at the time of its discovery, wagered me a new hat that it was one.

EXPLORATIONS.

Successive years of active exploration, hunting, and road or trail mak-ing in the park, have rendered the most of it, west of Yellowstone Lake and its Grand Cañon, so familiar that *research* is perhaps now more ap-propriate than exploration, for our observations therein. Still, there are now many localities of considerable area, as much of Mounts Ste-phens and Dunraven ranges, as little known as before Washburn scaled the peak which bears his name. Traversing such regions are truly ex-plorations, prominent among which, of this season, is that of the Madison Divide, in search of a pass to avoid the cliffs near Mary's Lake. Those to the south were explored last year and found utterly impracticable, although a depression observed this year in the crest of the range to the north afforded a hope that a pass might be discovered there. The long, open, but unsafe valley of hot springs and sulphur vents on the head of Alum Creek was traced to its connection with a branch of the Rocky Fork of the East Fire Hole River, and one mountain feeder of this, through an elevated divide, to the seething brimstone basin of Violet Creek, and another to a similar repellent sulphur region overlooking the Norris Geyser Basin and Fork of the Gibbon, and thence down the Rocky Fork to our camp on the East Fire Hole, and the effort there abandoned. Although this exploration failed in its main object, it led to the discovery and opening of a fine bridle-path route from above the mouth of Rocky Fork, through the earthquake region to the Paint Pots on the main road, which proved a good 20 miles saving of distance for our couriers and pack-trains from the headquarters to our camp on the Mary's Lake route. It also greatly extended our knowledge of the fire holes in those regions, and afforded proof positive that a band of bison wintered there, at an ele-vation of nearly 9,000 feet. Much was also learned of the broad elevated timbered plateau of Elephant's Back, and its extension above the Nat-ural Bridge; and exceedingly interesting knowledge was obtained of the apparently most recent shattering of the earth's crust, with still yawning impassable vents and lava overflow in this region of the Park, upon the various branches of the Blacktail, skirting the Great Cañon of the Yellowstone between the mouth of Crevice Gulch, via the head of Pleasant Valley to Tower Creek. By far the most extensive, interesting, and valuable exploration of the season is that in connection with, or continuation of, that of Governor Hoyt and Colonel Mason, in the Sierra Shoshone and main Rocky Ranges, during twenty-six days of continuous and arduous cliff and cañon climbing among the snowy lava-capped crests of a region of as wild chaotic grandeur, and as little known or understood as any other in the United States, if not indeed in North America. A journal of the transactions of each day was regularly kept, water-courses mappèd, prominent mountain peaks sketched, passes noted, and the weather and elevations recorded at least three times a day. Only the size and purposes of this report preclude its publication entire herein, but the preceding descriptions of the Two Ocean and other passes, the subjoined record of weather and elevations (the former accurate, and the latter, for want of reliability in the readings of the aneroid barometer, approximate only), the mountains and streams as shown upon the map, will be found tolerably correct, and it is hoped will prove of sufficient interest to encourage the attention of scientists better prepared and outfitted than myself to do this wonderful region justice.

HEADQUARTERS OF THE PARK.

One of my first and most important official duties in the Park was the search for a location for its headquarters, which should combine, in the fullest degree, nearness and accessibility throughout the year, through one of the two main entrances to the park; to the nearest permanent settlements of whites and a military post, remoteness from routes inviting Indian raids, and a proper site for defense therefrom, for ourselves, saddle and other animals, good pasturage, water, and timber, as well as accessibility to the other prominent points of interest in the Park. The want of any public funds in 1877 prevented other than exploration of routes to and throughout portions of the park (cut short by a severe injury at Tower Falls, just in advance of Chief Joseph's Nez Percé Indian raid), and the publication of a report.

The Bannock Indian raid of 1878 rendered unsafe the construction of public buildings or the retention of public property in the Park during the following winter, but the road constructed that year, connecting the two entrances from the Mammoth Hot Springs to the Forks of the Fire Holes, together with its value to myself in making other improvements, to the Hayden geological explorations, and to Generals Miles and Brisbin, in their military operations, confirmed my opinion, in which these gentlemen concurred, that the Mammoth Hot Springs was then, beyond question, the proper location for the headquarters of the Park. The buildings of hewn timber were mainly constructed in 1879, upon a commanding site for outlook and safety, the main one being surmounted by a loopholed gun-turret for defense from Indians. Subsequent explorations and improvements in the Park have justified the selection, alike of the location and of the building site. These are well shown, with the adjacent cliff fences to our large and valuable pasturage, in the frontispiece of the Park Reports of 1879 and of 1880; and the buildings as they now are in the frontispiece of this report.

As explained in my report of 1879, there was found at the Mammoth Hot Springs only one building site not overlooked by others, which one, besides its position commanding every locality within rifle range, was desirable from its gradual slopes and accessibility from the Upper Terraces, as well as direct connection with the matchless pastures and meadows beside and below it. The elevation of this building site from actual measurements is found to be : Above the Cedar Grove toward the Great Terraces southwesterly, 84 feet; above the Liberty Cap northwesterly, 152 feet; above the Little Meadows southeasterly, 226 feet; while towards the northeast the descent by terraces is nearly continuous for over a mile in distance, and fully 1,000 feet in descent, to the Great Medicinal Springs in the cañon of the Gardiner. Although so elevated and commanding a site for observation or defense, a depression down its least elevated side affords an excellent roadway upon each side of it, and between them a convenient location for a reservoir of warm water, which has proved alike useful for ourselves, for our animals, and for the purpose of irrigating our garden, especially for its protection during frosty nights. This hill was originally a sage-brush dotted, grassy mound, having a few dwarf firs and cedars upon it, and with a regular supply of cold water in a natural depression for a reservoir near the house, which might, with little expense, soon be shaded and screened by an evergreen grove, and with a supply of the terrace building water, furnish bathing rooms and ornament any desired portion of the slopes with peerless bathing pools like the ancient ones fronting it. For convenience, for symmetry, as well as for safety from gales, the main building, 40 by 18 feet, was built upon a stone foundation embracing our cellar, with one lean-to wing, 22 by 13 feet, for office and small bedroom, another, 25 by 13, for family sitting-room and bedroom, and a rear kitchen, 18 by 13 (see cut of ground plan, Fig. 2).

All these, together with the main edifice, are built of well-hewn logs notched and spiked or pinned together log by log as laid up, the attic portion of the wings thus sustaining the upper story of the main building, which is surmounted by an octagon turret 9 feet in diameter and 10 feet high from a solid foundation of timbers upon the plates, the upper ends of the well hewn and fitted timbers of which, extending above the roof, are loop-holed for rifles. From the evident infrequency of injury by lightning in the park, I ventured upon an additional mode of sustaining the building during wind-storms, as well as for providing a substantial flag-staff. This was done by planting a fine liberty-pole firmly in the rocky foundation of the building around which it was constructed, and to which it was firmly attached by several heavy iron bands, which allowed for the natural settling of the building, and thence extended through the center of the shingle-roofed octagon turret, above which, 53 feet from the main floor of the building, are the globe and flag-pulley. Altogether it is a sightly, substantial, and commodious building for a headquarters, only needing ceilings in the lower and partitions and ceilings in the upper story—both of which are high and airy—for its completion. The other buildings are an earth-roofed barn of hewn timber, 32 by 18, one end of the lower story of which is for a stable, and the other is an open front room for our wagons, &c. From the adjoining large and substantial corral, one gateway leads to the lane in front, and the other to the pasture in the rear.

A large, warm, and convenient hennery in the hillside near the barn has proved less valuable than was anticipated from the ceaseless destruction of our domestic fowls by the ever pestiferous mountain skunks. In the cedar grove near the old corral and reservoir is our round-log, earth-roofed blacksmith shop, 20 by 14. Amid the cedars at the foot of the cliffs is our rude partitioned bath-house, and at a proper distance in the rear of our main building is a commodious out-house. A large wire-screened box in the cool, sheltered nook at the north angles of the building is found valuable as a protection from blow-flies upon the elk meat and venison, which seldom taints at any season of the year.

All these buildings are detached and isolated beyond danger of ordinary fires, the constant fear of which induced the recent construction of a fire, frost, and burglar-proof vault, 12 by 16 feet, in the face of the dug-

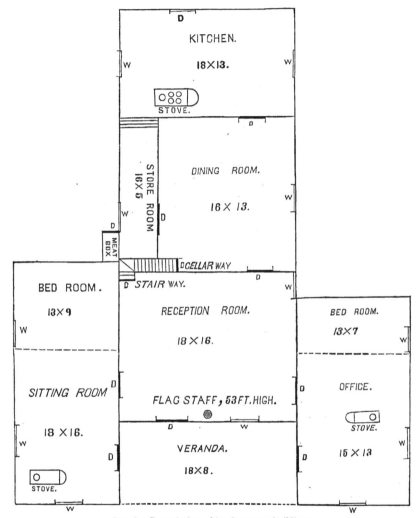

FIG. 2.—Ground plan of headquarters building.

way in rear of the main building, as a provision, tool, and outfit storehouse. These buildings have proved convenient, well adapted for the public purposes, and, saving improvement in a supply of good cold water, which is still more difficult to obtain in the Fire Hole regions, ample and substantial enough for headquarters, until the rapidly-approaching railroads demonstrate the necessity of others, and the proper location for them. This will admit of all the funds which may be appropriated for the park being expended for its protection, and the construction of roads, bridges, and other necessary improvements. Meanwhile some of the finest locations in the Fire Hole regions should be reserved from sale or leasehold to persons or railroad companies, from which to select a site for the head-quarters of the superintendent or his assistant, as may then be deemed best; it being evident that after the completion of railroads to the Mammoth Hot Springs, and to the forks of the Fire Holes, a leading officer of the park, with adequate buildings, will be a necessity at each of these places.

MAMMOTH HOT SPRINGS.

The characteristic tendency of these springs to dwindle or fail in one place and burst forth in another not remote has been very marked during this season in both location and power. We have been compelled to culvert the outlet of a hot spring which burst forth in our road at the foot of the Devil's Thumb during the past winter, and which is still active, while the springs near McCartney's Hotel dwindled until it was neces-sary to remove his bath-houses, and then burst forth anew in full power. The water, which has heretofore been too hot for comfort at our bath-house, was this year too cold for that purpose, or to properly protect our garden by irrigation during frosty nights, while a new pool, too far below it for use, is a veritable boiling caldron, and similar changes are observable on all of the terraces. Not only this, but the aggregate quan-tity of water upon these terraces is evidently diminishing, while that of the Hot Creek, which is fully 1,000 feet below, near the McGuirk Spring, on the Gardiner River, is surely increasing, but is not now of the terrace building, but of the medicinal class of springs.

LIBERTY CAP.

The suggestions contained in my report of 1880, in reference to recoat-ing this famous extinct geyser cone by a jet of water from the terrace building, Mammoth Hot Springs, having been approved, I decided to practically test whether these waters deposit at the orifice of a tube by evaporation only, or by deposition its whole length. For this purpose the open-ended double-barrels of a shot-gun were placed where a cur-rent of the hot water in a boiling spring passed steadily through them to the muzzle end, which alone protruded from the scalloped border. Repeated trials, resulting in filling the barrels within a week, demon-strated that *these* springs do certainly fill a tube by deposition the whole length, and not by evaporation at its exposed extremity, as had been believed. Hence the negotiation for the purchase of gas-pipe was aban-doned and water conveyed in troughs made for the purpose to the Dev-il's Thumb, and with perfect success, it having been covered and enlarged by a coating of beautiful white geyserite. The flow of water is now dis-continued for the purpose of learning if this coating will endure the frosts of winter; and if so, it only requires about 300 feet of scaffolding from 25 to 45 feet high to conduct the water from the Devil's Thumb to the Liberty Cap, and by building around the base, filling the fractures, and recoating it to thus preserve and beautify one of the unique marvels of the Park.

LAWS RELATING TO THE PARK.

All the enactments by Congress in reference to the vast regions in-cluded in the Yellowstone National Park may be found, first, in two brief sections approved March 1, 1872, dedicating it as a national health and pleasure resort, and placing it absolutely under the appropriate control of the Department of the Interior; and, second, by virtue of the annual appropriations during the past four years, aggregating up to July 1, 1882, the sum of $50,000, to enable the honorable Secretary of the Interior to protect, preserve, and improve it. For a knowledge of the enactment, see appendix marked A and regarding the second, or a proper showing of the management of these funds, and the manner and results of the expenditure, reference is made to the annual reports of the honorable Secretary, containing those of the superintendent thereof. The park has been wholly managed without the aid of the civil or military authorities of those regions, (save occasional assistance by the latter in repelling hostile Indians) under rules and regulations as prescribed by the honorable Secretary of the Interior, somewhat modified by experience. Those now in force will be found in appendix marked B. While under these rules and management, as fully shown in these reports, and included in maps, plates, &c., much has been peacefully accomplished (so far as the whites are concerned), in both protection and improvement of the park, it is believed that additional provisions by Congress, by the council of Wyoming Territory, or by both of them, are necessary, as well as the proposed organization of a county of Wyoming, with a seat of justice near enough to insure legal co-operation and assistance in the management of the park, as it is neither desirable nor in accordance with the spirit of our institutions, or of our people, to continue the control of so vast a region, teeming with people from nearly every land, by mere moral suasion, occasionally sustained by more potent appeals from the muzzles of Winchester rifles.

GUIDES OF THE PARK.

From the statements and letters of persons who visited or attempted to visit the park, I have no more doubt that many persons have been deceived, and have suffered from the greed, ignorance, or inefficiency of persons in the adjacent regions professing to be able to properly convey or guide tourists to and throughout the park, than of my utter inability or power to prevent such impositions. In addition to my present purpose of publishing a complete and accurate map and guide book of the park, for use during the coming season, I may add that I know of many good, honorable men, thoroughly acquainted with the park, its approaches, and its wonders, who will neither deface nor destroy guide-boards or represent that the park is destitute of roads, and that valiant guides and an arsenal of arms are indispensable to reach or safely visit its marvels or swindle or neglect those employing and confiding in them. If, in compliance with the earnest request of such persons now pending, I should adopt the policy of granting licenses, operative during good behavior, each season, which should cost such persons only the expense of badges, license, and record, holding each in a degree interested and responsible for the prevention of fires and acts of vandalism, and observance of the other rules and regulations for the management of the park, by the parties in their charge, I cannot doubt the result would be far more beneficial to the park, and its visitors, than pleasant to the superintendent, from the machinations of those whom he might deem unworthy to receive or retain such a license.

SUGGESTIONS REGARDING A POLICE FORCE FOR THE PARK.

As will be found in the interesting report of the gamekeeper, his experience and observations, as such, leads to the conclusion that an officer especially for the protection of game is not necessary in the park, but rather that there should be a small force of men, hired by the superintendent for their known worth, and subject to discharge for cause, or some of them, at the close of each season, in which opinion, from years of experience, I heartily concur. Selected as these men would be, from those hired as laborers, the hope of winning promotion to this more attractive and responsible duty would prove alike an incentive to win and faithfulness to retain it; and I am unaware of any other plan promising such efficient assistants in the indispensable protection of game, prevention of fire and vandalism, keeping regular records of the weather, and geyser eruptions, and in general asisting the worthy, and restraining the unworthy visitors of the various geyser basins, as well as for patrol for like purposes and for seeing to the roads and bridle-paths. There has not occurred a serious fire in the park since the Bannock raid from the camp fire of any of our laborers or of the mountaineers; but such is the inexcusable carelessness of many tourists, that without great watchfulness disastrous conflagrations, utterly impossible to check when once started, may yet destroy the matchless evergreen groves, and cover much of the park with impassable fallen timber.

Since writing the above, I am in receipt of a synopsis of Lieutenant-General Sheridan's report to the Adjutant-General of the Army, of his recent tour through the National Park, and his views and suggestions in reference thereto. Owing to his entrance to the Park from Fort Custer and and Clarke's Fork pass, he crossed the Yellowstone River at its forks, while Governor Hoyt, Colonel Mason, and myself were crossing it at the foot of the lake, some 40 miles above, en route to the Stinkingwater, and hence I failed in a desired interview with him, but it is with great pleasure that I acknowledge, in behalf of the park, my obligations to him for authorizing the reconnaissance of Colonel Mason, Captain Stanton, and Lieutenant Steever, and also to the first of these gentlemen for the courtesy (and assistance when needed) which has ever characterized the military officers with whom I have met in the park, as well as for a manuscript synopsis of his past season's explorations; and to the last two officers for their tables of odometer measurements—the first ever made of any of our roads or bridle-paths within the park. From the route taken by General Sheridan, via Mount Washburn bridle-path, he was unable to visit our headquarters or main line of improvements then completed in the park, but the tone of his remarks upon the magnitude of the National Park, the difficulties of its protection and improvement, the inadequacy of the means heretofore provided therefor, and his views as to a remedy, evince alike his intuitive comprehension of a subject or a region, and his military stand-point of view in the management of them.

REGISTERING THE NAMES OF TOURISTS.

The register of the names of tourists at the headquarters, is so incomplete regarding those known to have been there as not to justify publication; that of Job's Hotel, at the Mammoth Hot Springs, has not been received, but that of the Marshall House at the Forks of the Fire Holes, the remaining residence within the Park, although very incomplete, is published, hoping that it may prompt more attention to the matter hereafter by all parties. Various suggestions have been made as to the best mode of obtaining the names of all visitors to the park, one of which is the establishment of a gate and keeper at each of the two main entrances to the Park to compel registration of names, residence, and dates, which, besides the cost of the gates and keepers, would, I fear, prove unreliable to intercept or prevent false registration by those desirous of avoiding it, and which certainly would be incomplete, as the mountaineer tourists will hereafter enter the Park from nearly all quarters. Besides it may appear to many so like unjustifiable annoyance, that I incline to leave to time, the approaching railroads, increase of hotels, and wishes of the constantly multiplying number of tourists, for a solution of this matter.

REGISTER OF VISITORS.

Copy of the register of the Marshall Hotel at the forks of the Fire Hole rivers, Yellowstone National Park, from June 27 to August 25, 1881.

Date.	Name.	Residence.
1881.		
June 27	Charles R. Brodix	Bloomington, Ill.
July 14	Patrick Walsh	Virginia City, Mont.
25	James R. Johnson	Prickley Pear, Mont.
25	C. L. Dahler	Virginia City, Mont.
25	N. I. Davis	Do.
25	John McManus	Kirkville, Mont.
25	E. Panabacker	Do.
26	James R. Johnson	Prickley Pear, Mont.
26	Francis Collins	Pittsburgh, Pa.
26	R. K. Cooper	Silver City, Mont.
28	William Collins and wife	Glasgow, Scotland.
29	I. W. Thorne	Helena, Mont.
29	I. L. Mears	Wicks, Mont.
29	E. H. Metcalf	Do.
Aug. 3	George Huston and two men	Clarke's Fork, Mont.
3	E. Panabacker	Do.
3	R. Pearsall Smith	Philadelphia, Pa.
3	Hannah Whithall Smith	Do.
3	Mary W. Smith	Do.
3	Alys W. Smith	Do.
3	David Scull, jr	Do.
3	Edward L. Scull	Do.
3	William E. Scull	Do.
3	I. Tucker Burr	Boston, Mass.
3	Winthrop M. Burr	Do.
3	William S. Mills	Wilmington, Del.
3	Bond V. Thomas	Baltimore, Md.
6	Justice W. Strong	Washington, D. C.
6	Senator John Sherman	Ohio.
6	Senator Benjamin Harrison	Indiana.
6	Gov. B. T. Potts	Helena, Mont.
6	Albert Bierstadt, artist	New York.
6	P. W. Norris, sup- rintendent	National Park.
6	Judge W. H. Miller	Indiana.
6	Gen. Thomas A. Sharpe	Do.
6	E. Sharpe	Do.
6	Alfred M. Hoyt	New York.
6	E. W. Knight	Helena, Mont.
8	Dr. D. S. Snively	U. S. Army.
8	Lieut. W. D. Huntington	Do.
8	Miss H. D. Huntington	Fort Ellis, Mont.
8	Miss A. J. McKay	New York.
8	Z. H. Daniels	Bozeman, Mont.
9	Judge William Gaslin	Kearney, Neb.
9	Com. T. T. Oakes	New York.
9	James Gamble	San Francisco, Cal.
9	I. H. Hammond	Evanston, Ind.
9	Edward Stone	Walla Walla, Wash. Ter.
9	Gen. L. S. Willson	Bozeman, Mont.
9	L. W. Langhorne	Do.
9	E. L. Fridley	Do.
9	George Ashe	Do.
9	R. McDonald	Do.
9	I. V. Bogart	Do.
9	Fred. de Gamga	Senegambia.
9	Commodore Bell	Do.
13	Wm. F. Bowers	Boston, Mass.
14	Gov. John Hoyt	Cheyenne, Wyo.
14	Col. J. W. Mason	Fort Washakie Wyo.
14	Capt. John Cummings	Do.
14	P. W. Norris, superintendent	National Park.
14	Keppler Hoyt	Cheyenne, Wyo.
14	J. A. Mason	Fort Washakie, Wyo.
14	Harry Yount, gamekeeper	National Park.
14	G. W. Watkins	Towanda, Pa.

Copy of the register of the Marshall Hotel, &c.—Continued.

Date.	Name.	Residence.
1881.		
Aug. 14	Frank Grounds	Bozeman, Mont.
15	W. H. Young, sr	Butte, Mont.
15	W. H. Young, jr., and wife	Do.
15	H. Romsbush	Do.
15	I. G. Corrie	Do.
15	Miss Lizzie Astde	Do.
15	Francis Frances	England.
15	V. W. Benzing	New York.
15	Lieut. Edgar Z. Steever	U. S. Army.
20	John F. Forbes	Butte, Mont.
20	W. T. Hawley	Do.
	J. V. Long	Do.
	John Farwell	Do.
	Geo. N. Givin	Do.
	I. F. Rumsey	Chicago, Ill.
	Prof. W. I. Marshall	Fitchburg, Mass.
20	C. R. Hermon	Saint Lou's, Mo.
20	W. R. Larcey	Bozeman, Mont.
24	Walter Cooper	Do.
24	Geo. W. Wakefield	Do.
24	R. Koch	Do.
24	Fred. La Hare	Do.
25	Thomas Dennison	Cold Bluff, Pa.
25	W. C. Cady	New London, Ct.
25	P. W. Lytle	Oakdale, Pa.
25	A. J. Fisk	Helena, Mont.
25	Henry Cannon	Do.
25	G. R. Melten	Do.
25	W. E. Sanders	Do.
25	John Porter	Do.
25	C. A. Brown	Virginia City, Mont.

FISHES OF THE PARK.

Suckers, catfish, and the bony white mountain herring, abound in the Yellowstone River and some of the lakes, but far the larger portion of all the fishes found in the known waters of the Park are trout. These appear to me to be of many different varieties. Several of these are peculiar to a certain lake, as the red-gilled and red-finned trout of the famous Lake Abundance, at the head of Slough Creek, which has an area of less than a square mile. These trout are very beautiful as well as palatable when in flesh—then weighing nearly a pound each—but they often so overstock the lake as to become as voracious as sharks and too poor for food.

TROUT LAKE.

This noted lake or pond is situated about two miles above the famous Soda Butte, and is wholly supplied by a snow-fed rivulet less than a mile in length and only a good pace in width, and drained by another of similar dimensions, each having impassible cascades within one-fourth of that distance from it; and yet in this little isolated pond are found incredible numbers of one of the largest, most beautiful, and delicious trout of the entire mountain regions. In the spawning season of each year they literally fill the inlet, and can be caught in countless numbers. From my journal of June 3, 1881, I quote as follows:

Wishing a supply of trout for our men in the Gardiner Cañon, Rowland, Cutler, and myself rode to Trout Lake, and, after pacing around and sketching it, with brush and sods I slightly obstructed its inlet near the mouth. Within eight minutes thereafter the boys had driven down so many trout that we had upon the bank all that were desired, and the obstruction was removed, allowing the water to run off, and within three minutes thereafter we counted out 82 of them from 10 to 26 inches in length. Of these, 42 of the larger ones, aggregating over 100 pounds, were retained for use, 30 of the smaller ones returned to the lake unharmed, and the remaining 10 were, together with a fine supply of spawn, distributed in Longfellow's and other adjacent ponds, which, although as large, and some of them apparently as favorable for fish as the Trout Lake, are wholly destitute of them.

Although the boys declared this was not a favorable morning for trout, and they do doubtless often make greater hauls, still this is as large a fish story as I dare publish, and qualify even this with the statement that the pond is unusually full of weeds and grass, and the food supply of insects so abundant that the fish are not reduced in numbers by the rod as in many other ponds, and hence the incredible number in its small inlet during the spawning season. Trout varying greatly in size and appearance are found in the snow-fed rivulet branches of Alum Creek and other streams, whose waters are too hot and too full of minerals to sustain ordinary life.

FISHES OF THE YELLOWSTONE LAKE.

The only variety of fishes known to inhabit this great lake is the yellowish speckled salmon trout, which are usually found of from 15 to 25 inches in length. These are proverbial alike for their taking the hook so near boiling pools at various localities along the shore line that they may with ease be cooked in them upon the line without the fisherman changing his position, and for the large number of them being infested with long slender white worms. The proportion of them thus diseased has increased from something over one half in 1870 until all are apparently infested, as I have neither seen nor heard of one of the countless numbers caught this season which was clear of these parasites; and so many were dying along the shores, and so great the quantity of weeds with adherent sacks of yellowish-green jelly, that they drift in lines—sometimes in small windrows—along the shore. Not only this, but it is the opinion of those the best acquainted with this lake that its waters are more discolored with these weeds and less pure than formerly. What degree of connection, if any, these various peculiarities hold to each other, is only conjectural, but to assist in an investigation I have sent the skin, a portion of the meat, entrails, and worms of one of these trout in a bottle of alcohol, and some of the sprigs of this weed and sacks, as well as porous yellowish stone tubes of some worm or insect which are found in abundance along the bank of the lake, to Prof. S. F. Baird, director of the Smithsonian and National Museum, and United States Commissioner of Fish and Fisheries. It has been suggested that some other more voracious fish might exterminate the trout and stock the lake, but whether the latter would prove any more exempt from the parasites, evidently depends upon whether the disease is peculiar to the trout, or to the lake; the evidence now known favoring the latter theory, as trout thus diseased are found only in this lake, or in waters so connected with it as to indicate that they frequent it. Thus, I have no knowledge of a worm-infested trout having been found in the Yellowstone River below the Great Falls; although many of the trout there are apparently of the same species with those of the lake, and presumably some of them may, at some period of their growth, have safely passed the falls; or, waiving this theory, trout of the same variety are never, as I am aware, found thus infested in the numerous mountain feeders of the Snake branch of the Columbia, which so interlock in the Two Ocean and other passes, that there is strong probability that the trout, like the waters, do actually intermingle, and would become diseased also did the cause pertain to the fish and not to the lake. These are the facts so far as now known, and the subject being one of both scientific and practical importance, in connection with a lake of about 200 miles shore line, and far the largest of its elevation upon the globe, I earnestly invite a thorough investigation and pledge all the assistance in my power to render it as complete as possible.

Sufficient time has not elapsed as yet to determine the results of my experiments in stocking various lakes this season with trout, but I propose to extend the effort in larger lakes, like Shoshone and Lewis, and shall report progress from time to time.

In view of the paucity of species of fishes in the Park, it is my earnest intention during my next season's explorations to endeavor to find suitable waters in which to attempt the culture of carp, a subject which is now engrossing quite a large share of the attention of those interested in cheap and nutritious food fishes.

HISTORY OF THE PARK.

Since embracing in this chapter of my report of 1880 my previous private and official publications regarding the aboriginal inhabitants and early white rovers of the park, the accumulation of material from exploration, research, and the narratives of trappers and miners in these regions, as well as the perusal of rare publications in the east, is deemed sufficient to justify a synopsis of them herein. This material, for brevity and clearness, is arranged as follows:

First. Traces of a people who inhabited these regions prior to their occupancy by the present race of Indians.

Second. Remains of Indians.

Third. Evidence of early white trappers.

Fourth. Narrations of prospecting white men before the Washburn exploring expedition of 1870.

Fifth. Explorations of this year.

TRACES OF A SUPPOSED PREHISTORIC PEOPLE.

These consist mainly of utensils, weapons, and implements not now or known to have ever been used by the present race of Indians in or adjacent to the park. Also, rude stone-heap drive-ways for game, which I have recently found therein, or adjacent thereto, some of which are here represented.

Notes regarding ollas, vessels of stone, &c., found in the Yellowstone National Park in 1881.

Fig. 3. Fragment of steatite vessel, size restored about as follows: Inches.

Greatest diameter --- 11
Height externally --- 10
Depth of vessel inside -- 8
Breadth of rim -- 1

Much thicker in the bottom, pecked into oval shape outside. No evidence of fire, but some pestle marks in the bottom of the cavity. Found upon the surface in the Upper Madison Cañon.

Fig. 4. Fragment of steatite vessel, size restored: Inches.

Greatest diameter --- 8
Height externally --- 10
Depth of vessel inside -- 9
Breadth of rim -- ½

Very uniform throughout and finely finished, but not polished or ornamented; showing very evident *fire marks*. Found outside up, nearly covered with washings from the volcanic cliffs, together with various rude stone lance-heads, knives, and scrapers, in the remains of ancient

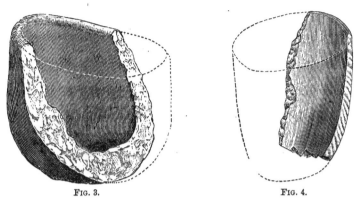

FIG. 3.   FIG. 4.

camp-fires disclosed by the recent burning of the forest border of the upper end of Pleasant Valley, on the right of where our road enters it from the cliffs.

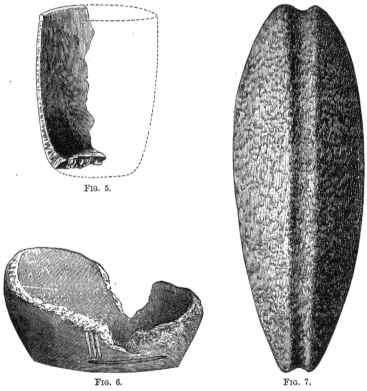

FIG. 5.

FIG. 6.   FIG. 7.

Fig. 5. Fragment of a steatite vessel, size restored:

	Inches.
Greatest diameter	7
Height externally	12
Depth of vessel inside	10
Breadth of rim	$\frac{5}{8}$

Very uniform, well finished outside, but showing much evidence of fine tool-marks inside. Found upon the surface of the mines at the head of Soda Butte.

Fig. 6. Soapstone or very soft steatite vessel, fragment :　　　Inches.

Greatest diameter	5
Smallest diameter	$3\frac{1}{2}$
Height externally	$2\frac{3}{4}$
Depth of vessel inside	2
Breadth of rim	$\frac{1}{2}$

Well finished inside and out, with flat bottom. No evidence of fire; found with fragments of pottery and rude lance-heads at an ancient camp on the eroding bank of the Blacktail Creek. It may be mentioned that these steatite vessels are the first found between the Atlantic and Pacific coasts, and are entirely different in form from those found in either direction.

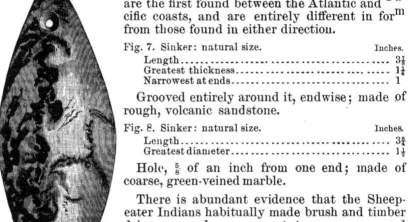

FIG. 8.

Fig. 7. Sinker: natural size.　　　Inches.

Length	$3\frac{7}{8}$
Greatest thickness	$1\frac{1}{4}$
Narrowest at ends	1

Grooved entirely around it, endwise; made of rough, volcanic sandstone.

Fig. 8. Sinker: natural size.　　　Inches.

Length	$3\frac{3}{4}$
Greatest diameter	$1\frac{1}{4}$

Hole, $\frac{5}{8}$ of an inch from one end; made of coarse, green-veined marble.

There is abundant evidence that the Sheep-eater Indians habitually made brush and timber driveways and arrow coverts to secure game, and little to show that their progenitors or predecessors ever found timber so scarce in the park as to require driveways to be made of long lines of small stone-heaps such as are found; and this year I traced and sketched from the commencement of the open valley of the Yellowstone, upon the borders of the park below the mouth of Gardiner River, through the Bottler Park and the Gate of the Mountains, to the open plains, a distance of fully 60 miles. As this is mainly outside of the park, and the exploration exhausted none of the funds appropriated therefor, the report and numerous sketches of these stone-heaps, the cliffs over which at least the buffalo were driven, traces of bone-heaps, rude stone foundations of dwellings, together with their burial cairns, mining-shafts, and the tools, ornaments, and weapons obtained from them, will be published elsewhere in due time.

Fig. 9 is a representation of a line of rude stone heaps, probably intended as a driveway for game over the cliffs upon the banks of the Yellowstone. The stone circles shown are evidently the foundations of very ancient dwellings, as the stones like those of the driveway are about one-half covered with accumulated débris. This sketch may be considered as typical of others, many of which are much larger.

I will only here add that there is proof positive of the early and long occupancy of these mountain parks and valleys by a people whose tools, weapons, burial cairns, and habits were very unlike those of the red Indians, and who were the makers of the steatite vessels, &c., we discovered; but whether they were a branch of the cliff-dwellers of the cañons of the Colorado, progenitors of the Sheepeaters, or both or

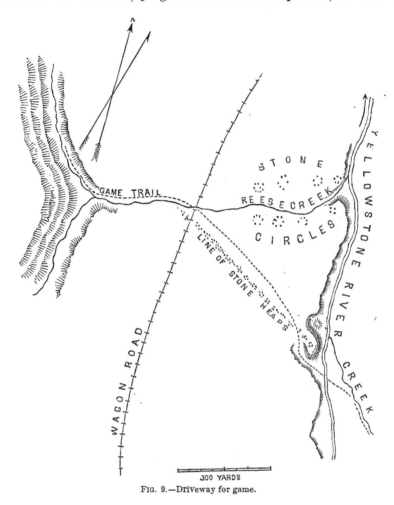

FIG. 9.—Driveway for game.

neither, are questions better understood with exploration and research in that direction, which I have commenced and hope may be continued.

INDIAN REMAINS.

These are, first, of the various kinds usually found in regions until recently only occasionally visited rather than inhabited by the nomadic hunter tribes, such as trails, lodge-poles, brush wick-e-ups, peeling of timber, and rude storm or timber wind-brakes upon commanding sites or narrow passes, for observation, ambush, or for protection from their enemies or the elements, as well as rude stone axes, or flint or obsidian knives, lance and arrow heads and scrapers; and, second, those pertaining to the timid Sheepeater occupants, such as remains of camp-fires in the secluded glens or cañons, and occasionally in caves or niches in the cliffs, for shelter from the storms, or seclusion or defense from their enemies; timber driveways for animals to some well-chosen place for arrow-covert ambush and slaughter, and notably an occasional circular breastwork of timber or stone, or, as is common, partly of each, as to the real builders of which, and the purposes for which constructed, opinions differ. Four of these were discovered during this season, viz, one beside our camp, in a grove north of the crossing of Willow Creek, some three miles below Mary's Lake, which was seen by Hon. John Sherman and party, including the artist Bierstadt, who sketched it. It is about thirty feet long by twenty wide, and constructed of fragments of logs, stumps, poles, and stones, with ingenuity and skill proverbial

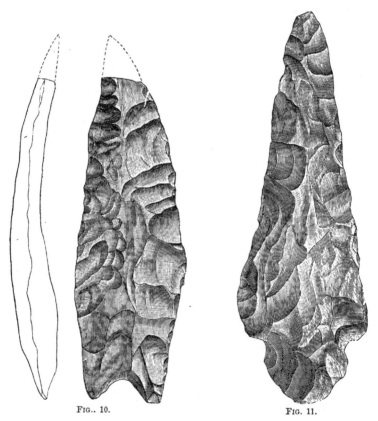

Fig.. 10.   Fig. 11.

to the beaver; nearly weather, wind, and bullet proof; about breast high, which is certainly less than when built, and situated, as usual, in a wind-fall then screened by a thicket of small pines, which are now large enough for bridge or building timber. A similar one was found upon the Stinkingwater side of the pass, which I discovered this season, in the Sierra Shoshone range, east of the Yellowstone Lake; another near Bridger's Lake, and the newest one on a small branch of Barlow Fork of Snake River. Although these and some of those previously found do not appear older than some of the evidences of white men, others certainly do, but none of them in any part of their construction as yet known show an iron ax or hatchet hack upon them, and very few and faint marks of even stone tools or weapons. There is usually little evidence of a door or gateway, and none of a roof, but abundant proof of a central fire, and usually of bones fractured lengthwise for the extraction of the marrow, as practiced by many barbaric peoples.

While these constructions much resemble a Blackfoot Indian fort, the infrequency of the visits of these Indians to a region of few horses, the utter lack of marks of hatchets, which they have long possessed and al-ways use, dispels this theory, and, as Sheep-eater wick-e-ups and pole coverts

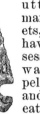

FIG. 13.   FIG. 14.   FIG. 15.

FIG. 12.

under low and heavily branched trees are common for summer use, and on cliffs for winter, the only remaining and most probable theory is that these are really winter lodges of the Sheep-eaters in the thicket borders of warm, sheltered valleys, where the abundant timber of the decaying wind-falls, in which they are always found, could be liberally used in an inclosure so large as to not take fire, while it was a great protection against the cold, even if, without being wholly or in part covered with the skins of animals, and as necessary against the prowling wolf and wol-verine in winter, the ferocious grizzly in spring, or human foes on occasions. Two Shoshone Indian scouts and guides accompanied the exploring expedition of Governor Hoyt and Colonel Mason during the past season, one of whom, We-saw, had accompanied Cap-tain Jones in his explorations of 1873. As arranged with Governor Hoyt at Mary's Lake, I en route tried the Nez Percé ford, and deeming it then barely passi-ble for those well acquainted with the channel both sides of the island, posted a notice so informing him at the Mud Vol-cano, and then ascended the river to its head, laid out some work for my laborers, constructed a raft, and crossed in time to intercept the governor and party, who, after visiting the Great Falls, had crossed at the Nez Percé ford under the skillful guidance of We-saw, although this was his first visit since the one with Captain Jones, eight years before, and the channel had meanwhile changed materially. While the rest of the party were camped at Concretion Cove, We-saw and myself went over to the Jones trail on the Pelican, and thence followed it as near as possible for large areas of timber fallen since his visit, to the entrance of his pass of the Sierra Shoshone range, and in order to avoid this timber selected a route via the Hot Springs feeder of Turbid Lake. During this and other occasions I could not fail to ad-mire the intuitive accuracy of his judgment as to Jones's and other routes, even where no trace was visible, and in various conversations as well as comparisons of our daily sketches, which each regularly kept in his own style, obtained much valuable information. I found him an old but remarkably intelligent Indian, and so accurate in his sketches that I could readily trace them, although they were destitute of the point of compass, date, or word of explanation; and yet in that, as in all else,

FIG. 17.

FIG. 16.

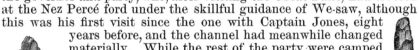

FIG. 18.

FIG. 19.

he manifested the true Indian character, which, like their farms, *is all long and no wide*, i. e., a keenness of perception rather than a broad or general comprehension of a subject, or even a region. Hence, although a person skilled in Indian sketching could by his map or sketch easily follow them through a long journey in all its turns and windings, neither one or both of them could therefrom make a general map of the region or of the relative positions of various mountains or other portions of the route, even approximately, and this is in fact the main difficulty with the maps and journals of white rovers also. We-saw states that he had neither knowledge nor tradition of any permanent occupants of the Park save the timid Sheepeaters, his account of whom is embraced in the history of them. He said that his people (Shoshones), the Bannocks, and Crows occasionally visited the Yellowstone Lake and river portions of the Park, but very seldom the geyser regions, which he declared were "*heap heap bad*," and never wintered there, as white men sometimes did with horses; that he had made several trips before the one with Captain Jones, one of which was, as I understood him, to assist some friends who had intermarried with the Sheepeaters to leave the Park after the great small-pox visitation some twenty years ago. Among the most recent as well as the most interesting of Indian remains are those heretofore reported of the rudely fortified camp of Chief Joseph and his Nez Percés in 1877. Of these, the corral east of Mary's Lake, corral and small breastwork between the Mud Volcano and the river, and others upon Pelican and Cache Creeks, and their dugways in descending into the cañons of Crandall and Clarke's Forks, possess peculiar historic interest. Figs. 10 to 24, inclusive, represent natural size scrapers, knives, lance or spear heads, perforators, and arrow-heads chipped from black obsidian. These were found in various places, such as caverns, driveways, or at the foot of cliffs over which animals had been driven to slaughter, and are typical of a collection of over two hundred such specimens collected this season.

EARLY WHITE ROVERS IN THE PARK—JOHN COULTER.

Since the publication in my report of 1880 of a reference to the trip of the Indian gauntlet-running Coulter across the National Park, in 1808 or 1809, I have, through the kindness of General O. M. Poe, Corps of Engineers, U. S. A., obtained a trace of the prior wanderings of this famous mountaineer, of which, as well as of the map exhibiting them, I had no previous knowledge. This map is contained in the first of three rare volumes now in the military library of the War Department, Washington, and is an English reprint in 1815 of the journals of Lewis and Clarke to the head of the Missouri River, and across the continent to the Pacific and return, during the years 1804-'05 and '06. The portion of this map showing the routes of these explorers, is remarkably accurate and the rest of it a fair representation of what was then known of those regions; but, as that was a medley of fact and fiction, of truth and romance, gleaned from the narratives of three centuries of Spanish rovers from Mexico, two of French missionaries or traders from Canada, and the more recent and more accurate accounts, English or Americans, between them, far the most valuable fact shown was the existence of an elevated snowy fountain-head and point of divergence for nearly all of the mighty rivers of central North America, while the real or relative location of the upper portion of all save those visited by Lewis and Clarke, as there shown, are at best only approximate, and are now known to be mainly erroneous. A knowledge of these facts is alike necessary to properly estimate the truth and the errors of this map, and especially those portions of the country shown to have been visited by Coulter in 1807. After his honorable discharge, as stated by Lewis and Clarke, in 1806, near the mouth of the Yellowstone, he ascended it to Prior's Creek, a southern branch of the Yellowstone, between the Clarke's Fork and the Bighorn, where he probably wintered, and, as shown by the map, the next year traversed the famous Prior Gap to the Clarke's Fork, which he ascended nearly to its head, and thence crossed the Amethyst Mountain to the main Yellowstone River, and that at the best ford ͺupon it. This is the famous Nez Percé ford at the Mud Volcano, the location of which is accurately shown under the name of Hot Brimstone Spring. But, most strangely, neither the Great Falls of the Yellowstone, 10 miles below, nor the lake, 8 miles above, are represented; but the river is correctly shown as a very wide one, not only to where the foot of

FIG. 20:

the lake really is, but also incorrectly throughout its length, and the locating of one of the fingers to and as being the outlet of Eustus Lake, which he reached by crossing the main divide of the Rocky Mountains without knowing it. This is pardonable, as from the peculiar situation in the mountains of the lake he called Eustus (evidently Shoshone) Lake which was mistaken by Professor Hayden and others as Atlantic and not Pacific waters, only they thought it drained into the Madison, and Coulter supposed it drained into the Yellowstone, while it is in fact the head of one fork of the Snake River of the Columbia, although from its size (12 miles long) Coulter deemed it the large lake at the head of the Yellowstone, of which he must have heard.

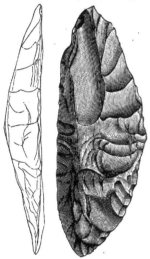

From this lake his route seems to have been through or near Two Ocean and Tog-wo-tee Passes to lake Riddle, which, though far too large, is, from its location and drainage into the

FIG. 21.

Upper Bighorn River, probably the Great Hot Spring at the present Fort Washakie, near the Wind River Shoshone Indian Agency. He thence traversed the mountains to Coulter's Fork of the Rio del Norte, as he naturally deemed it, discharging into the Gulf of Mexico, while it is in fact the Green River of the Colorado, of Major Powell's Grand Cañon to the Gulf of California. In traversing the South Pass he crossed the Continental Divide probably for the sixth time, without knowing it, to the Platte, which he calls the Rio de la Plata, and thence across

FIG. 22.

the mountains and Bighorn River, through fossil regions, to the Salt (really South) Fork of the Stinkingwater, to the great Stinking Spring near the forks, and hence the name which, Indian-like, does not signify *the river of stinking water,* but the river which passes or is near the *stinking water.* From this are two trails, evidently a division of the unknown party, of probably whites and Crow or Shoshone Indians, one of which makes a cut-off to the outward trail on Clarke's Fork, and the other through much of the Bighorn region and a Gap Creek Pass to Prior's, and presumably the Yellowstone River at its mouth. This map shows a band of Snake (Shoshone) Indians, called *Yeppe,* of 1,000 souls, at the location of Pelican Creek and a valley, which, together with the Yellowstone Lake, as above shown, were neither visited nor correctly represented, but with little doubt this is the band to which the Shoshone Indian We-saw referred, as well as some of the evidences at Concretion Cove, in the preceding section

FIG. 23.

upon Indian Remains. I have devoted unusual space to this matter, which I think is of great interest, as being the earliest known record of white men in any portion of the National Park, and is nearly as valuable for what is erroneously as well as for that which is correctly represented, from being a compilation by the highest authority of all that was at that period known of those vast mountain regions, and in no way conflicts with the account of the death of Potts, during Coulter's gauntlet-running expedition upon the Jefferson, or his return through the Park, as that was a subsequent

FIG. 24.

expedition, and probably unknown to Lewis and Clarke at the time of their first publication of their journals, of which this English edition was mainly a reprint.

RECORDS OF THE EARLIEST WHITE MEN FOUND IN THE PARK.

The next earliest evidence of white men in the Park, of which I have any knowledge, was discovered by myself at our camp in the little glen, where our bridle-path from the lake makes its last approach to the rapids, one-fourth of a mile above the upper falls. About breast-high upon the west side of a smooth pine tree, about 20 inches in diameter, were found, legibly carved through the bark, and not materially obliterated by overgrowth or decay, in Roman capitals and Arabic numerals, the following record:

FIG. 25.

The camp was soon in excitement, the members of our party developing a marked diversity of opinion as to the real age of the record, the most experienced favoring the theory that it was really made at the date as represented. Upon the other side of this tree were several small wooden pins, such as were formerly often used in fastening wolverine and other skins while drying (of the actual age of which there was no clew further than that they were very old), but there were certain hatchet hacks near the record, which all agreed were of the same age, and that by cutting them out and counting the layers or annual growths the question should be decided. This was done, and although the layers were unusually thin, they were mainly distinct, and, in the minds of all present, decisive; and as this was upon the 29th day of July, it was only one month short of sixty-two years since some unknown white man had there stood and recorded his visit to the roaring rapids of the "Mystic River," before the birth of any of the band of stalwart but bronzed and grizzled mountaineers who were then grouped around it. This is all which was then or subsequently learned, or perhaps ever will be, of the maker of the record, unless a search which is now in progress results in proving these initials to be those of some early rover of these regions. Prominent among these was a famous Hudson Bay trapper, named Ross, whose grave I have often seen (the last time in going to the Bighole battlefield for the bones of Lieutenant Bradley, in 1879) where he was long since killed by the Blackfeet Indians in Ross's Hole— as parks were then called—at the head of the Ross Fork of Bitter Root branch of the Hell Gate, in Montana, and which was named after him; as was also, perhaps, the branch of Snake River in Idaho, where the Shoshone Indian Agency is situated. The "R" in the record suggests, rather than proves, identity, which, if established, would be important, as confirming the reality of the legendary visits of the Hudson Bay trappers to the Park at that early day. Thorough search of the grove in which this tree is situated only proved that it was a long-abandoned camping ground. Our intelligent, observant mountaineer comrade, Phelps, upon this, as upon previous and subsequent occasions, favored the oldest date claimed by any one, of the traces of men, and, as usual, proved to be correct.

The narrowest place of the Yellowstone River of which I have any knowledge below the lake is between our camp of the Glen and the Upper Falls; and upon the eastern rock, just above the latter, I had often seen a medium-sized stump, which Phelps declared was cut by himself when returning with two or three comrades from James Stuart's fruitless Big Horn expedition of 1864, or seventeen years before this time, and that if we would cross the river he would show us the ruins of their camp-fire also. This we soon after did with a raft (as the river was then too high to cross as I have frequently done later in the season), in taking the measurements of the river for a future bridge, and where claimed by Phelps found the charred fire-brands of the camp, tent-poles, and even picket-pins for the lariats of the horses, intact, and, saving at the surface of the ground, but little decayed; in fact, the hatchet hacks upon all of the poles, including the ends of the pins, although of pine, in the ground, were uniformly clear and distinct. In company with this comrade I subsequently visited a scaffold for drying meat, at a ford of the Pelican, which I had often at a glance in passing deemed four or five years old, which he accurately described before reaching, and at once recognized as one of their camps of 1864, although he had not in the intervening time visited the vicinity. From the appearance of these and many other camps which were subsequently visited with him, or recognized by his description at various places in the mountains (including a pass near that of Two Ocean), and which I thus particularly mention as being, save those of Captain De Lacy hereafter mentioned, the oldest traces of white men in the Park, of which we have positive data, I learned to judge of the relative age of certain marks, which, from signs hard to explain, were unmistakably recognized as the traces of unknown white men. In addition to the old loop-holed log ruin near the brink of the Grand Cañon below Mount Washburn; the cache of old Hudson Bay marten traps, near Obsidian Cliffs; decaying stumps of foot-logs over Hellroaring and Crevice streams, and other evidences of early white men, heretofore mentioned in my reports, I saw many during this season's explorations, a few only of which will be here noticed.

In the grove-girt border to the small lake back of Concretion Cove of Yellowstone Lake are the traces of very old tree and brush shelters for horses, larger and differently formed from those of Indians, and the numerous decaying bones of horses, proving that they died probably by starvation during some severe winter, or, as is less likely, were killed by the Indians in an attack before carrying the camp, (as they were not at that day properly armed with guns), for they would certainly have saved and not slaughtered them thereafter. Stumps of trees, re-mains of old camps, and the fragments of a rough dugout canoe, prove that white trappers long since frequented the famous willow swamps around the mouths of the Upper Yellowstone and the Beaver Dam Creek. Our first noon camp in ascending the east side of the Yellow-stone above the lake was purposely made where Harry Yount found a human cranium in 1878. This skull we failed to find, but we utilized some of the wood cut and split, but not corded, by white men so long ago that, though the upper cross-sticks were apparently not decayed, they were dried into curvature from the heart and seams in the well-known manner of timber unearthed from peat bogs or beaver dams, and were easily broken over the knee by a sudden pressure of the hands upon the ends; also one end of a long pole for camp purposes, thrust through the fork of a pine, was there much overgrown. This camp was made near the eastern edge of a then new wind-fall of timber, as shown by the fragments of logs chopped, extending from the river to a lovely lawn skirting the towering cliffs; a well-chosen place for defense, or for secretion, unless betrayed by the presence of horses. A little distance above the camp are the stumps of trees cut and one of the logs not used in the construction of a raft. This wind-fall is now overgrown by trees, certainly not less than fifty or sixty years old; and the skull, fire-wood, raft log, and other circumstances indicate that a party of white men were attacked, and, after loss of horses, at least some of them hastily left their camp and attempted to escape by descending the river. Just south of the trail between the South Creek and the summit of the Two Ocean Pass is one standing and several fallen posts, and some poles of what may have been a very large oblong square tent, or more probably a conical lodge, as the appearance of the notches in the top of these posts, to sustain strong ridge and plate poles, seem to indicate that it was inclosed with skins and not canvas. But as the notches in the top of the posts were unquestionably made by white men, it was probably constructed for some grand council between the early trappers and the Indians, of which we have no other record or tradition than these decay-ing remnants.

The deep broad, and often branched bridle-paths up the Pelican Creek have usually been attributed to the thousands of horses of the retreat-ing hostile Nez Percés or Bannocks and their white pursuers in 1877 and 1878, but this year I followed heavy trails from Camp Lovely, near the open pass from the South Fork of Pelican Creek, down an unknown branch (which these Indians did not follow) to the East Fork of the Yellowstone, finding constant evidences of camps and other distinctly recognized traces of white men, made long years before the miles of burned and fallen timber—now much decayed—caused the abandonment of the route.

In closing this interesting subject it is only added that to tradition and slight published records I find abundant wide-spread, and, to my mind, conclusive evidence that white men frequented these regions nearly or quite from the visit of Coulter in 1807 until the waning of the fur trade after the discovery of gold in California, and in a lesser degree continuously thereafter. What portion of these rovers were trusty trappers and what hiding outlaws will never be known. Nor is it material to history, as the interest of each conduced to a successful concealment from the public of a knowledge of the cliff and snow girt parks and valleys of the National Park, fully two generations after the surrounding regions, some of which are fully as inaccessible, were well known, correctly mapped, and published to the world.

WHITE PROSPECTING MINERS.

The dwindling of placer mines in California, and their discovery else where, greatly increased the numbers of the worthy prospecting successors of these roving trappers, and these were joined during the war of the rebellion by many deserters from the Union and Confederate armies, and by refugees from the devastated borders between them, and bold men from elsewhere; who preferred fighting Indians in the West to white men in the East, being mostly armed with long-range breech-loading rifles. Scarce since the days of the Pilgrims of the Cross, and the wild crusade of the mailed warriors of Europe for the sacred tomb in Palestine, has the world witnessed an onset more wide-spread, daring, or resistless than that of the grim gold-seeking pilgrims to Wyoming, Idaho, and Montana. Streaming from the East, organized, often broken up and reorganized upon the plains, under Bridger, Bozeman, or other daring leaders, they, with wagon trains, pack trains, on horseback or afoot, collectively or separately, fought their way through the Cheyenne, the Sioux, and other of the fiercest fighting Indian nations of the plains, with bull-boat, raft, or wagon, afoot or on horseback, forded, ferried, or swam the mighty rivers, and in bands, in squads, or alone, poured a resistless stream through nearly every mountain pass, yawning gulch, and dangerous cañon, to all the main parks and valleys from the Platte to the Columbia.

Of some of these parties and pilgrims we have knowledge, but doubtless many prospectors have traversed these regions, visited portions of the park prior to 1870, but as they were seeking mines, and not marvels, and better skilled in fighting Indians than in reporting discoveries, the little known of them is being learned from their own recent publications, or by interviews with those of them still living, the list embracing many of the wealthiest and worthiest citizens of these regions, the narratives of some of which are added.

On page 113 of the first volume of the History of Montana is found the commencement of a very interesting narrative by Capt. W. W. De Lacy, now and long a prominent and esteemed surveyor and engineer of Montana, of the wanderings of himself and party of prospecting friends during the latter part of 1863. Leaving Alder Gulch, now Virginia City, in Montana, August 3, they crossed the main divide at Red Rock Creek, and proceeded thence, via Camas, Market Lake, and the forks of Snake River, and through the broken regions of East Fork, so graphically described in Irving's Astoria and Bonneville, reaching Jackson's Lake, at the very foot of the towering Tetons. Here the party divided, one portion returning via Lewis Lake and the Fire Hole and Madison Rivers to Virginia City, while Captain De Lacy, with twenty-six men, missed Lewis Lake, but discovered and skirted a lake which was very properly called after their leader, De Lacy. This was named and published in maps for years before Professor Hayden or any of his men saw it; and some of them, for some unknown cause, gave it the name of Shoshone, which, though a fitting record of the name of the Indians who frequented it, is still in my view a gross injustice to its worthy discoverer, as, even if my interpretation of Coulter's visit in 1807 is correct, it was then unknown. From this lake De Lacy and party crossed the main divide of the Rocky Mountains to the East Fire Hole River, which they descended to the forks, and down the main Madison, through its upper cañon, then across the North Fork and through mountain defiles to the head of the west branch of the Gallatin Fork of the Missouri. The above narrative, the high character of its writer, his mainly correct description of the regions visited, and the traces which I have found of this party, proves alike its entire truthfulness, and the injustice of changing the name of De Lacy's Lake; and fearing it is now too late to restore the proper name to it, I have, as a small token of deserved justice, named the stream and park crossed by our trail above the Shoshone Lake after their discoverer.

The journey of G. H. Phelps and comrades connected with the armed expedition of James Stuart early in the spring of 1864, to the Bighorn regions, for the purpose of avenging the slaughter of some, and the terrible sufferings of the rest, of his party, in 1863; failing to find the Indians, they broke up into prospecting parties, that of Phelps wandering through the mountains to the Sweet Water, through the South Pass to Green River, then to the Buffalo Fork of Snake River, crossing the main divide in the pass near Two Ocean, which, as before stated, I recognized from his description, and attached his name. Thence they descended to Bridger's Lake, crossed the Upper Yellowstone, and continued upon the east side of it, as well as of the lake and lower river, past Pelican Creek and the falls, as before shown, to the trail of another party of white men, which they followed to Emigrant Gulch, near the Gate of the Mountains.

From a well-informed and truthful mountaineer, named Adam Miller, I learned the history of this party. In the spring of 1864, H. W. Wayant, now a leading citizen of Silver City, Idaho, William Hamilton, and other prospectors, to the number of forty men, with saddle horses, pack train, and outfit, ascended the east side of the Yellowstone from the Gate of the Mountains to Emigrant, Bear, and Crevice Gulches, forks of the Yellowstone, East Fork, and Soda Butte; thence over the western foothills of Mount Norris to the bluffs upon the south side of Cache Creek, where their horses were all stolen by some unknown Indians, but their only two donkeys would not stampede, and remained with them. Here the party broke up; Wayant, Harrison, and ten others, with one jack, and what he and the men could carry, ascended Cache Creek to Crandall Creek, Clarke's Fork, Heart Mountain, thence by way of Index Peak and the Soda Butte returned to the cache made by the other party of what they could not carry, aided by their donkey, from where set afoot, and hence called Cache Creek. They then crossed the East Fork, scaled the Amethyst Mountain, forded the main Yellowstone, at Tower Falls, and thence returned via the mouth of Gardiner River, Cinnabar, and Cañon Creek, where I saw traces of them in 1870, to Alder Gulch, now Virginia City, Montana. Meanwhile the other party had returned, and some of them assisted in planting the mining camps of Crevice, Bear, and Emigrant.

Later in the same season George Huston and party ascended the main Fire Hole River, and from the marvelous eruption of the Giantess and other geysers, and the suffocating fumes of brimstone, fearing they were nearing the infernal regions, hastily decamped. These, with the visit of Frederick Bottler, and H. Sprague, Barronette, and others mentioned in preceding reports, are the most important of those as yet known, until 1870.

Upon a pine tree, below the confluence of the North Fork of the Stinking Water and the creek which we ascended to the new pass, is plainly and recently carved as follows:

FIG. 26.

Evidently showing that some one, on the 5th day of some month, the name of which commences with A, failed in an effort to ascend the stream, and so informed some person or party, who would then have known the date and circumstances. This record may have been left by a member of A Company, Fifth Infantry, this company having been with General Miles in the Bannock campaign of 1878, or the famous mountaineer and guide, Yellowstone Kelley, may have carved it.

A square pen of logs, with a huge dead-fall at its only entrance, found on Orange Creek, is certainly a white man's bear-trap, and like many other traces is of uncertain date, and not of sufficient interest for further notice.

INDIAN TREATIES.

The first white visitors to the National Park found the timid, harmless Sheepeater Indians the only permanent occupants of it; their nearest neighbors, the Bannocks, Shoshones, and Mountain Crows, its most frequent visitors; and the occasional prowlers therein, the rapacious Blackfeet and Sioux, robbers of their race, and the early white trappers or these regions. Decimation by war and disease, with the occupancy of intervening regions by whites, guarantee future safety from the Blackfeet; a nearly impassable mountain range and a cordon of military posts and armed ranchmen, from the Sioux.

SHEEPEATERS, BANNOCKS, AND SHOSHONES.

The recent sale of the National Park and adjacent regions by these Indians insures future freedom from any save small horse-stealing bands of these tribes also. To prevent these forays, in council at their agency on Ross Fork of Snake River, in Idaho, and in Ruby Valley, in Montana, early in 1880, I obtained a solemn pledge from them to not thereafter go east of Henry's Lake, in Montana, or north of Hart Lake, in Wyoming, to which, as stated on page 3 of my report of 1880, they faithfully adhered. This pledge was renewed at Ross' Fork when I was en route from Washington this year, and has again been sacredly observed. Unable to visit the Lemhi Agency of these tribes, by letter I represented the matter, and sent printed copies of the rules and regulations for the management of the Park to Maj. E. C. Stone, their agent, who, in reply under date of May 26, stated that, after mature deliberation in council, he felt justified in pledging that the Indians of his agency would not thereafter enter the Park. The only known disregard of this pledge was by a band of three lodges of hunters upon the North Branch of the Madison, which was promptly reported and checked, and is not likely to occur again.

MOUNTAIN CROWS.

These Indians, numbering about 3,000, have as a tribe never been hostile to the whites, but often their valuable allies in conflicts with others, and though beset with their proverbial craving for horses, without a special observance of brands or collar-marks, besides some minor failings too prevalent with other races also, they have by the sale of much of their lands, and granting the right of way for a railroad through the remainder, proven that, although in common with their race they may be the guilty possessors of a valuable region desired by the all-absorbing white man, still they are not intentional obstructors in the pathway of progress.

As shown in my preceding reports, sustained by memorials of the officers and other leading citizens of Montana, and proven by the records, the following facts are established:

First. No portion of the northern or western watershed of the Yellowstone Range, between the Gate of the Mountains and the borders of Wyoming, including a three-mile strip of the Yellowstone National Park in Montana, was ever occupied, owned, or even claimed by the Crows, save only as being embraced in the then unknown boundaries of their reservation as set off in 1868.

Second. In 1864, or four years prior to the cession of this land to the Crows, the Sheepeater Indians, owners of Emigrant, Bear, and Crevice Gulches, had been dispossessed by the white miners, who have since constantly occupied portions and controlled all of it, with the full knowledge and acquiescence of the Crows.

Third. Upon the discovery of mines upon the northeastern watershed of the said Yellowstone Range, below the Gate of the Mountains, which had always been owned and occupied by the Crows, they promptly sold the entire range, embracing alike that occupied by the miners and that by themselves, including the old agency, buildings and improvements, as well as valuable agricultural lands, and have for many months allowed white men to occupy it, although, by the delay of Congress to appropriate the funds, they are still without one dollar of pay therefor. Besides this, they have, as before stated, shown their peaceful and progressive tendencies by promptly granting for a mere nominal sum a very liberal right of way along the whole river front of the remainder of their reservation for a railroad artery of civilization. Meanwhile, mines, mills, ranches, and the site and buildings of at least one village (Emigrant, or Chico), with a United States post-office, are, in the absence of all lawful organization or protection, held only by actual possession, without legal right of transfer or even improvement, which are alike indispensable to attract capital for the development of a most promising mining and agricultural region. Hence, in justice and good faith alike to the white man and to the Indian—to the Crow who surrendered a region without remuneration, and to the miner who holds it without title; to the race dwindling away for want of civilization for the means which are their due of obtaining it; to the poor but dauntless path-finding prospector of boundless hidden wealth for the race of resistless destiny sure reward for its discovery and development, and for the peaceful adjustment and legal occupancy of a border of the Wonder Land of earth, and the safety of those who may visit, improve, or occupy it, do I urge, through the active influence of the department, the speedy appropriation by Congress of the means to cancel treaty obligations by paying this confiding people for a valuable region long since peacefully surrendered.

As the hostile incursion of Chief Joseph and his Nez Percés in 1877 was the armed migration of a people, anomalous in all its features, and impossible to ever again occur, with the peaceful adjustment of these Crow difficulties closes all claims or danger of Indians in any portion of the Park, and with it the necessity or semblance of an excuse for tourists to traverse it stalking arsenals of long-range rifles and other weapons, merely to slaughter or frighten away the dwindling remnant of our noblest animals, which it should be the pride as it is the duty of our American people to here preserve from threatened extinction.

HOODOO OR GOBLIN LAND.

A trail was opened this season upon a nearer route than that followed last year, and some new discoveries made around the base of Mount Norris, upon Cache Creek, and thence in nearly a direct line to and beyond the Hoodoo Mountain to Mason's Creek, at the head of the Great Stinkingwater Cañon, near the forks of which is a yawning cañon bordered by unearthly goblin forms as hideous as any conjured in wildest dreams.

C. M. Stephens accompanied me from the Mud Volcano to Clarke's Fork, with his transit for the purpose of taking daily and nightly observations; but although in early September we were terribly annoyed by fogs and storms, from the summit of Mount Chittenden we, protected by overcoats and gloves, through occasional rifts in the fog-clouds, got fair views of the Yellowstone Lake and Pelican Creek regions, but not of the Hoodoo, and upon the latter during the entire day of September 6 we remained, amid chilling fogs which were ascending from the melting snows in all the adjacent valleys, standing behind our monument of last year with compass and field-glass, ready to catch every glimpse of sunshine or opening in the shifting mists below or about us, and at various times obtained fair bearings of most of the leading points of interest, save Index Peak, which was not visible during the entire day. We proposed renewing our observations the next day, and then descend the Middle Fork of Crandall Creek to an open grassy plateau which we had plainly seen from the mountain, but a few miles distant upon Clarke's Fork, to the northeast. But the terrific snow-storm, which had kept us in a clump of fir trees at our camp of last year during much of the 4th and all of the 5th, recommenced with such fury that we hastily descended along our new trail about 30 miles to the gamekeeper's cabin on the Soda Butte, where the weather was warm and pleasant, with little snow. Determined to complete the exploration, leaving our pack-animals and outfit, we ascended the Soda Butte 20 miles to Clarke's Fork Mines, and spent the rest of the day in viewing the pass to Clarke's Fork and a route to Crandall's Creek for the morrow's effort. With the dawn came a snow-storm so furious that we yielded to the inevitable, and pressing through the storm, which as we descended decreased to no snow and a bright sunset at the cabin that night. The next day I returned through mingled snow and sunshine, 35 miles, reaching our headquarters on the eve of September 10, which I had only visited once for a few moments since the morning of July 1.

METEOROLOGICAL RECORD.

Weather record, kept by P. W. Norris, during the exploration of the Sierra Shoshone and a portion of the Rocky ranges.

[* Indicates approximate elevation only.—P. W. N.]

Date.	Camp.	Location.	Time.	Elevation.	Ther.	Weather.	Wind.
1881.				*Feet.*	°		
Aug. 16	1	Two miles below Mary's Lake...	7 a. m...	7, 500	51	Cloudy	SW.
16	N.	Mud Volcano..................	Noon ...	7, 725	65	Rainy	SW.
16	2	West side of the foot of Yellowstone Lake.	6 p. m...	7, 738	61	Clear	S.
17	2	......do....................	6 a. m...	7, 738	41	.. do	N.
17	2	......do...	Noon ...	7, 738	70	....do	SW.
17	2	......do...................	6 p. m...	7, 738	62	Fair	SW.
18	2	......do...................	6 a. m...	7, 738	42	....do	NE.
18	3	Concretion Cove, on Yellowstone Lake.	Noon ...	7, 738	68	Cloudy	SW.
18	3	......do...................	Sunset..	7, 738	51	Windy	S.
19	3	......do............	Sunrise .	7, 738	40	Cloudy	N.
19	N.	Jones' Pass of the Sierra Shoshone Range.	Noon ...	9, 444	60	....do ..... :	SW.
19	4	Mouth of Jones' Creek, near Jones' Camp No. 36.	Sunset..	6, 683	54	Fair	SW.
20	4	.. ..do..............	Sunrise .	6, 683	34	Clear	S.
20	5	Jones' Camp No. 35, at head of the Grand Stinkingwater Cañon.	1 p. m ..	6, 319	85	....do	N.
20	5	......do...	Sunset..	6, 319	65	....do	SW.
21	5	......do..................	Sunrise .	6, 319	51	... do	SW.
21	N.	Snow field on Bald Mountain...	1 p. m ...	*10, 650	28	....do	SW.
21	5	Camp No. 5, on the Stinkingwater.	8 p. m...	6, 319	65	....do	SW.
22	5	......do..................	Sunrise .	6, 319	64	Slight shower	NW.
22	N.	Noon halt on the Norris Creek...	Noon ...	*7, 500	73	Clear	NW.
22	6	Forks of Norris Creek..........	Sunset..	*7, 812	50	....do	W.
23	6	.. ..do...	Sunrise .	*7, 812	31	....do	W.
23	N.	At pond and cascade in pass.....	Noon ...	*8, 476	43	... do	NW.
23	7	Forks of Clear Creek....	Sunset..	*7, 950	70	... do	W.
24	7	......do..................	Sunrise .	*7, 950	51	....do	W.
24	N.	Signal Point, Yellowstone Lake.	Noon ..	*7, 500	70	....do	NW.
24	8	Terrace at the head of the left finger of the Yellowstone Lake.	Sunset..	7, 800	65	.. do	S.
25	8	..... do..................	Sunrise .	7, 800	50	Thundershower.....	SW.
25	N.	At Old Hunters' Camp on the east side of the Upper Yellowstone.	Noon ...	7, 910	65	Clear	S.
25	9	Near Bridger's Lake	Sunset..	7, 950	58	Showery....	E.
26	9	.. ..do...	Sunrise .	7, 950	22	Clear	N.
26	10	Two Ocean Pass..............	Noon ..	8, 081	60	....do .:.....	SW.
26	10	......do..................	Sunset..	8, 081	51	....do	SW.
27	10	......do...	Sunrise .	8, 081	23	....do	NE.
27	N.	Continental Divide.............	10 a. m ..	*10, 100	45	....do	NE.
27	N.	Barlow Valley	1 p. m ..	8, 400	81	....do	SW.
27	11	Branch of Barlow River........	Sunset..	*8, 600	60	... do	N.
28	11	......do...	Sunrise .	*8, 000	21	....do	N.
28	N.	Summit of pass from branch of Yellowstone to one of Heart Lake.	Noon ...	8, 481	75	....do	SW
28	12	Head of Heart Lake............	Sunset..	7, 475	60	....do	S.
29	12	......do...	Sunrise .	7, 475	23	....do	N.
29	N.	Summit of Mount Sheridan	11 a. m ..	10, 386	60	....do	S.
29	12	Head of Heart Lake	Sunset..	7, 475	55	Cloudy.....	N.
30	12	..... do......	Sunrise .	7, 475	32	Snowy.....	N.
30	N.	Head of the thumb of the Yellowstone Lake.	Noon ...	7, 738	31	Snow squalls	SW.
30	13	Bridge Creek, near Natural Bridge.	Sunset..	7, 908	41	... do	SW.
31	13	......do	Sunrise .	7, 908	19	Clear	W.
31	14	Mud Volcano on Yellowstone River.	Noon ...	7, 725	60	Cloudy......	W.
31	14	......do.....................	Sunset..	7, 725	51	... do	W.
Sep. 1	14	......do..................	Sunrise .	7, 725	31	Hazy	SW.
1	N.	Pelican Creek..................	Noon ..	7, 800	58	....do	W.
1	15	Foot of Mount Chittenden	Sunset..	7, 850	43	do	W.
2	15	......do..................	Sunrise .	7, 850	22	Snow squalls	NW.
2	N.	Summit of Mount Chittenden...	9 a. m ...	10, 190	31	....do.	NW
2	N.	Pelican Creek.................	Noon ...	8, 000	48	....do.	NW

Weather record, kept by P. W. Norris, &c.—Continued.

Date.	Camp.	Location.	Time.	Elevation.	Ther.	Weather.	Wind.
1881.							
Sept. 2	16	Camp Lovely, in pass to the East Fork of the Yellowstone River.	Sunset..	*8,241	55	Clear	W.
3	16	.....do	Sunrise.	*8,241	14	....do.	W.
3	N.	East Fork Valley...............	Noon ..	7,180	55	Cloudy......	SE.
3	17	Three miles up Miller's Creek...	Sunset..	7,190	61	....do.	SE.
4	17	.....do	Sunrise.	7,190	50	....do.	SE.
4	N.	Head of Miller's Valley.........	11 a. m .	*7,350	48	....do.	NW.
4	18	Old camp, one mile from Hoodoo Mountain.	Sunset..	*8,490	35	Severe snow storm.	NE.
5	18	.....do......	Sunrise .	*8,490	30	....do.	NE.
5	18	.....do......	Noon ...	*8,490	35	....do.	NE.
5	18	.....do.............	Sunset..	*8,490	36	....do.	NE.
6	18	.....do......................	Sunrise .	*8,490	19	....do.	E.
6	N.	Summit of Hoodoo Mountain....	Noon ..	*10,700	21	Fogs........	NE.
6	19	Old camp, one mile from Hoodoo Mountain.	Sunset..	8,490	50	Clear	W.
7	19	.....do......................	Sunrise .	8,490	18	Snow squalls	E.
7	N.	East Fork of Yellowstone Valley	Noon ...	6,825	40	Clear	S.
7	20	Game keeper's cabin............	Sunset..	6,410	24	....do.	S.
8	20	.....do......................	Sunrise .	6,410	22	....do.	NW.
8	N.	Cook City or Clarke's Fork Mines	Noon....	7,590	36	Cloudy	NW.
8	21	Miller's Camp in the mines	Sunset..	8,425o	39	... do.	NW.
9	21	.....do...................	Sunrise	8,42	12	Severe snow storms.	N.
9	N.	Soda Butte......................	Noon...	6,500	26	...do.	N.
9	22	Game keeper's cabin............	Sunset..	6,410	26	Clear	SE.
10	22	.....do	Sunrise .	6,410	16	....do.	SE.
10	N.	Forks of the Yellowstone.......	Noon....	6,000	36	....do.	SW.
10	23	Mammoth Hot Springs..........	Sunset..	6,450	65	....do.	SW.

4 Y P

Meteorological record for the season kept at the Mammoth Hot Springs.

KEY.—S. R., sunrise; M., noon; S. S., sunset; Cl., clear; Cd., cloudy. No wind gauge.

[Latitude 44° 59′ north; longitude 110° 42′ west; elevation, 6,450 feet.]

January, 1881.

Date.	S. R.	M.	S. S.	Wind.	Snow or rain.	Sky.
1	24	36	26	NW.	....	Cd.
2	12	26	18	N.	....	Cd.
3	12	26	18	E.	....	Cd.
4	24	32	30	SE.	S.	Cd.
5	18	44	32	SE.	....	Cl.
6	10	30	28	S.	S.	Cd.
7	6	4	2	N.	S.	Cd.
8	14	6	0	N.	S.	Cd.
9	4	10	12	SE.	....	Cl.
10	16	20	22	S.	......	Cd.
11	28	34	36	SE.	S.	Cd.
12	34	34	36	SE.	S.	Cd.
13	10	22	18	W.	....	Cl.
14	18	32	34	SE.	S.	Cd.
15	28	50	40	SE.	....	Cl.
16	4	28	24	SE.	....	Cl.
17	1	27	25	SE.	....	Cl.
18	20	32	30	SE.	....	Cl.
19	30	48	20	NW.	....	Cd.
20	12	28	20	NW.	....	Cl.
21	18	28	26	SE.	....	Cl.
22	8	22	24	E.	....	Cl.
23	22	26	24	SE.	....	Cd.
24	18	38	20	SE.	....	Cd.
25	0	24	4	NE.	S.	Cd.
26	6	16	18	E.	....	Cd.
27	22	32	35	SE.	....	Cd.
28	32	30	34	SE.	....	Cd.
29	28	36	32	SE.	S.	Cd.
30	30	34	36	SE.	S.	Cd.
31	28	40	38	SE.	....	Cl.
	19	29	25		(Mean 24.)	

February, 1881.

Date.	S. R.	M.	S. S.	Wind.	Snow or rain.	Sky.
1	30	36	38	S.	R.	Cd.
2	39	41	42	SE.	R.	Cd.
3	42	44	44	S.	R.	Cd.
4	39	42	40	S.	R.	Cd.
5	32	51	39	E.	....	Cl.
6	29	50	39	....	....	Cl.
7	20	38	28	NW.	S.	Cd.
8	19	32	24	SE.	....	Cd.
9	8	30	22	NE.	....	Cd.
10	8	26	20	NE.	S.	Cd.
11	6	26	18	NE.	S.	Cd.
12	0	17	20	SE.	S.	Cd.
13	14	6	12	N.	S.	Cd.
14	4	14	2	W.	....	Cl.
15	2	16	16	SE.	....	Cl.
16	12	22	26	SE.	S.	Cd.
17	16	42	26	SE.	....	Cd.
18	9	29	20	SE.	....	Cl.
19	19	36	26	SE.	R. & S.	Cd.
20	9	24	22	SE.	....	Cd.
21	28	33	30	SE.	....	Cd.
22	32	39	36	W.	....	Cl.
23	32	40	34	S.	....	Cl.
24	32	42	38	S.	....	Cl.
25	34	32	34	SE.	S.	Cd.
26	26	39	30	SE.	....	Cl.
27	28	38	38	SE.	S.	Cd.
28	38	49	41	SE.	R.	Cd.
	22	33	29		(Mean 28.)	

March, 1881.

Date.	S. R.	M.	S. S.	Wind.	Snow or rain.	Sky.
1	32	38	34	SE.	....	Cl.
2	22	36	28	SE.	....	Cl.
3	26	38	32	SE.	....	Cl.
4	30	42	36	SE.	....	Cl.
5	22	44	35	SE.	....	Cl.
6	37	38	30	N.	....	Cd.
7	2	44	34	SE.	....	Cl.
8	18	40	33	SE.	....	Cl.
9	18	56	40	S.	....	Cl.
10	41	48	41	S.	....	Cd.
11	29	43	30	N.	....	Cl.
12	22	32	22	S.	....	Cl.
13	18	40	29	S.	....	Cl.
14	12	38	28	S.	....	Cl.
15	8	48	28	SE.	....	Cl.
16	11	46	32	SW.	....	Cl.
17	14	42	26	SE.	....	Cl.
18	25	38	29	N.	....	Cl.
19	26	52	32	SW.	....	Cl.
20	32	44	44	SE.	....	Cl.
21	28	59	40	SE.	....	Cl.
22	26	58	43	SW.	....	Cl.
23	30	50	54	SW.	....	Cl.
24	38	50	30	SW.	....	Cl.
25	28	48	45	SW.	....	Cl.
26	40	48	43	NE.	....	Cl.
27	40	48	43	NE.	....	Cl.
28	25	60	46	NE.	....	Cl.
29	32	66	50	SE.	....	Cl.
30	35	68	53	W.	....	Cl.
31	36	50	43	SE.	....	Cl.
	27	47	37		(Mean 37.)	

April, 1881.

Date.	S. R.	M.	S. S.	Wind.	Snow or rain.	Sky.
1	36	66	65	SE.	....	Cl.
2	36	53	41	SE.	....	Cl.
3	30	66	50	SE.	....	Cl.
4	34	68	54	SE.	....	Cl.
5	40	58	54	SE.	....	Cl.
6	44	50	40	SE.	R.	Cd.
7	24	35	30	NW.	....	Cd.
8	22	38	38	NW.	....	Cd.
9	24	32	24	NW.	S. & R.	Cd.
10	20	48	28	NW.	R.	Cd.
11	21	43	30	NW.	....	Cd.
12	18	42	36	NW.	....	Cl.
13	23	42	34	SE.	....	Cl.
14	34	48	46	SE.	....	Cl.
15	40	57	52	SE.	....	Cl.
16	42	62	56	SE.	....	Cl.
17	42	54	50	SE.	....	Cl.
18	44	50	45	SE.	....	Cl.
19	43	56	50	SE.	R.	Cd.
20	45	62	50	SE.	R.	Cd.
21	36	62	50	SE.	....	Cl.
22	40	49	40	SE.	R. ...	Cd.
23	32	44	40	SE.	....	Cd.
24	35	50	48	SE.	R. & S.	Cd.
25	40	48	42	SE.	....	Cl.
26	35	42	42	SE.	....	Cl.
27	37	50	47	SE.	....	Cl.
28	32	48	46	SE.	R.	Cd.
29	40	58	55	SE.	....	Cl.
30	40	56	50	SE.	....	Cl.
	34	51	44		(Mean 43.)	

Meteorological record for the season, &c.—Continued.

May, 1881.

Date	Temperature S.R.	M.	S.S.	Wind.	Snow or rain.	Sky.
1	40	56	50	S.	....	Cl.
2	40	60	50	SE.	....	Cl.
3	32	50	40	SW.	....	Cl.
4	40	60	56	S.	....	Cl.
5	40	60	56	S	....	Cl.
6	35	55	50	S.	....	Cd.
7	45	60	56	S.	....	Cl.
8	48	65	60	SE.	....	Cl.
9	45	66	60	SE.	....	Cl.
10	49	70	65	SE.	....	Cl.
11	40	60	56	SE.	....	Cl.
12	35	56	50	SE.	....	Cl.
13	35	45	40	SE.	....	Cd.
14	37	40	40	SE.	....	Cd.
15	35	48	45	SE.	....	Cd.
16	40	50	45	SE.	....	Cl.
17	45	60	57	SE.	....	Cl.
18	50	65	60	SE.	....	Cl.
19	50	66	60	SE.	....	Cl.
20	53	60	58	SE.	....	Cl.
21	50	62	60	SE.	....	Cl.
22	50	65	60	SE.	....	Cl.
23	50	65	60	SE.	....	Cl.
24	50	65	60	SE.	....	Cl.
25	45	60	58	SE.	....	Cl.
26	48	63	60	SE.	....	Cl.
27	50	65	60	SE.	....	Cl.
28	55	70	65	SE.	....	Cl.
29	50	65	63	SE.	....	Cl.
30	52	67	60	SE.	....	Cl.
31	54	68	62	SE.	....	Cl.
	45	60	56	(Mean 54.)		

June, 1881.

Date	Temperature S.R.	M.	S.S.	Wind.	Snow or rain.	Sky.
1	50	70	62	SE.	....	Cl.
2	48	83	74	SE.	....	Cl.
3	48	74	70	SE.	....	Cl.
4	42	65	60	SE.	....	Cl.
5	45	68	60	SE.	R.	Cd.
6	41	68	50	S.	....	Cd.
7	38	64	48	S.	....	Cd.
8	46	60	48	SE.	R.	Cd.
9	46	48	48	SE.	R.	Cd.
10	40	55	38	SE.	R.	Cd.
11	40	50	38	SE.	R.	Cd.
12	44	58	50	SE.	R.	Cd.
13	44	66	55	SE.	R.	Cd.
14	52	58	58	SE.	R.	Cd.
15	50	58	58	SE.	R.	Cd.
16	50	58	50	SW.	R.	Cd.
17	40	58	50	E.	....	Cl.
18	48	68	52	SE.	....	Cl.
19	55	70	53	SE.	....	Cl.
20	50	60	55	SE.	....	Cl.
21	45	60	55	SE.	....	Cl.
22	50	52	55	N.	....	Cl.
23	45	50	45	NW.	....	Cd.
24	50	80	45	S.	R.	R.
25	60	72	65	SE.	R.	R.
26	59	74	66	SE.	R.	R,
27	65	80	70	SE.	R.	R.
28	70	86	70	SE.	....	Cl.
29	70	82	70	SE.	....	Cl.
30	70	85	70	SE.	....	Cl.
	50	66	56	(Mean 57.)		

July, 1881.

Date	S.R.	M.	S.S.	Wind.	Snow or rain.	Sky.
1	53	98	84	(*)	(*)	(*)
2	70	88	84			
3	60	78	72			
4	55	75	70			
5	50	80	76			
6	68	82	78			
7	62	63	48			
8	33	62	60			
9	50	56	53			
10	47	68	55			
11	50	60	58			
12	53	78	74			
13	58	85	76			
14	60	80	74			
15	54	76	60			
16	56	76	60			
17	58	86	80			
18	68	90	82			
19	66	80	80			
20	54	64	59			
21	50	76	60			
22	50	83	70			
23	52	84	80			
24	52	80	58			
25	50	80	80			
26	54	74	68			
27	60	78	62			
28	53	82	70			
29	56	80	74			
30	54	84	80			
31	64	74	68			
	55	77	69	(Mean 67.)		

August, 1881.

Date	S.R.	M.	S.S.	Wind.	Snow or rain.	Sky.
1	44	72	65	(*)	(*)	(*)
2	46	80	72			
3	60	90	62			
4	60	92	65			
5	52	95	72			
6	60	94	87			
7	72	80	66			
8	60	76	74			
9	50	68	55			
10	60	74	70			
11	50	78	60			
12	46	88	74			
13	50	88	74			
14	60	74	70			
15	50	76	70			
16	52	75	70			
17	45	74	70			
18	48	80	65			
19	46	78	64			
20	47	85	60			
21	50	82	60			
22	50	85	60			
23	40	72	70			
24	50	86	58			
25	46	74	60			
26	54	86	74			
27	48	72	64			
28	40	76	54			
29	42	78	62			
30	40	64	60			
31	34	68	60			
	50	79	66	(Mean 65.)		

*No record of wind, rain, or sky.—C. M. S.

Meteorological record for the season, &c.—Continued.

Date	Temperature. S. R.	M.	S. S.	Wind.	Snow or rain.	Sky.	Date	Temperature. S. R.	M.	S. S.	Wind.	Snow or rain.	Sky.
	September, 1881.							**October, 1881.**					
1	50	62	56	(*)	(*)	(*)	1	40	50	48	SW.		
2	40	66	48				2	20	68	64	S.		
3	38	74	60				3	34	61	42	S.		
4	48	48	50				4	35	63	46	SE.		
5	34	46	38				5	38	68	50	E.		
6	30	58	58				6	38	41	40	E.		
7	24	46	42				7	32	38	42	NE.		
8	50	70	60				8	31	51	39	E.		
9	40	60	44				9	32	64	54	E.		
10	30	68	46				10	39	68	58	SE.		
11	44	74	64				11	34	40	32	N.		
12	40	80	74				12	10	35	18	SE.		
13	40	80	68				13	6	44	34	N.		
14	40	78	42				14	12	24	20	SE.		
15	34	86	70				15	9	30	28	NE.		
16	40	84	72				16	18	34	25	E.		
17	50	64	42				17	26	40	30	SE.		
18	36	74	62				18	31	42	30	SE		
19	36	60	58				19	36	53	40	SE.		
20	34	64	60			Cd.	20	38	54	40	W.		Cl.
21	40	60	56				21	34	47	42			Cl.
22	40	58	54				22	32	51	40	N.		Cl.
23	40	45	40				23	25	63	40	SE.		Cl.
24	30	48	40		S.		24	26	64	40	SE.		Cl.
25	40	55	35		R.		25	28	55	44	SE.		Cd.
26	26	48	32		S.	Cd.	26	32	48	42	SE.		Cd.
27	28	50	35			Cl.	27	30	50	40	SE.		Cl.
28	26	45	30		S.	Cd.	28	30	48	41	SE.	S.	Cd.
29	25	50	36		S.	Cd.	29	31	44	40	SE.		Cd.
30	28	40	35		S.	Cd.	30	34	44	32	SE.	R.	Cd.
							31	28	30	34	NW.		Cd.
	36	61	50	(Mean 49.)				29	49	39	(Mean 39.)		

November, 1881.

Date	Temperature. S. R.	M.	S. S.	Wind.	Sky.	Remarks.
1	30	35	31	NW.	Cd.	Rainy.
2	28	34	30	NW.	Cd.	
3	25	31	26	SE.	Cd.	Snow.
4	26	50	24	SE.	Cd.	Squally.
5	20	34	25	SE.	Cl.	
6	24	35	32	SE.	Cl.	
7	28	38	30	SE.	Cl.	
8	24	35	38	SE.	Cl.	
9	21	34	30	N.	Cl.	Windy.
10	20	31	24	N.	Cl.	Windy.
11	18	30	25	N.	Cd.	Snow squalls.
12	20	30	24	N.	Cd.	Snow squalls.
13	26	35	30	SE.	Cl.	
14	30	42	35	SE.	Cl.	
15	30	40	25	SE.	Cl.	
16	32	46	30	SE.	Cl.	
17	34	50	41	W.	Cd.	Rain.
18	34	46	38	W.	Cd.	Rain.
19	30	40	32	SW.	Cd.	Rain.
20	32	44	31	SW.	Cl.	
21	30	42	34	SW.	Cl.	
22	20	36	30	SE.	Cl.	
23	18	30	25	SE.	Cd.	
24	16	28	20	SE.	Cl.	
25	13	20	17	SE.	Cd.	Snow.
26	9	30	20	SE.	Cd.	Snow.
27	14	32	28	SE.	Cd.	Snow.
28	18	40	35	S.	Cl.	
29	19	40	30	S.	Cl.	
30	18	36	30	S.	Cl.	
	23	39	26	(Mean 29.)		

SULPHUR.

The demand for this mineral, for the purpose of preventing and curing the skin and hoof diseases of the sheep, increases, as does the animal, grazing upon the grassy slopes and terraced foot-hills of these mountain regions, where they are proving very profitable to their owners; and, as no refining of the substance is necessary for this and similar purposes, all thus needed by ranchmen could be readily obtained in the National Park, if they are allowed to do so, which has not been done further than to test its fitness and invite propositions. At the suggestion of the Hon. John Sherman, while we were visiting Sulphur Mountain during the past season, several excavations were made in the sulphur deposits of that and other localities, in order to learn something of their depth and quality. The uniform result was the finding of sulphur somewhat mixed with geyserite and other substances, in strata, or banding to where we were forced to desist by scalding hot sulphur water, or the stifling fumes arising from the deposit, at depths ranging from 3 to 6 feet from the surface. Specimens of these have been forwarded, with those of obsidian, geyserite, &c., to the National Museum for exhibition, as well as to obtain an opinion regarding their practical value. Although in this first search for beds of sulphur no heavy deposits cold enough to be worked were found, still I deem it far from conclusive evidence that none such exist, which may yet be found and profitably worked, if it be considered best to allow its being done. Hence, I suggest the propriety of allowing the search to be made by some responsible person or company, under a lease, allowing the mining and sale thereof of a limited quantity, and for a restricted length of time, and under such regulations as may be thought necessary and proper. While I do not in this desire to represent that any great revenue will immediately accrue to assist in the protection and improvement of the park, I see little danger of loss or injury in exploring some of its nearly countless sulphur deposits, but a certainty of obtaining many specimens of the fragile but beautiful sulphur crystals, and perhaps beds of commercial value, or knowledge of scientific interest.

PAINT-POTS.

This is a provincialism, or local phrase for the dwindled remnants of salses or mud geysers, which are difficult to describe or comprehend otherwise than by actual view of them.

Having in detail described the various kinds of geysers in my last year's report, I here only need to add that from the choking of the supply pipe, or fissure, to the regular intermittent Geyser, or from the bursting out of new ones, many of them dwindle into salses, with only an occasional eruption of their seething, foaming, muddy contents, and still dwindling in power, while increasing in their density and coloring, as well as the fetid smell, and nauseous, often noxious gasses escaping therefrom in spasmodic, hissing or gurgling throes or eruptions, become what are called paint pots. These are sometimes in gulches or basins commingled with or bordering the other kinds of geysers, but usually in more or less detached localities, each of which generally exhibits a preponderance of red, yellow, or other coloring characteristic of the predominant iron, sulphur, or other mineral substances of the basin, but in many of them are found closely and irregularly intermingled pools or pots of seething nauseous paint-like substances of nearly every color and shade of coloring known to the arts, and with a fineness of material and brilliancy of tinting seldom equalled in the productions of man. Although so brilliant, the colors of these paints are not permanent, but soon fade, and as the deposits are so numerous, accessible, and constantly accumulating, it is a question for scientific research to learn if the addition of lead or other minerals in proper proportions may not render these mineral paints practically valuable. There is direct evidence that the Indians used this paint liberally in adorning or besmearing their persons, their weapons, and their lodges. They also used a much more durable variety of red and yellow paint found in bands, layers, or detached masses, in the cliffs, a notable deposit of which was discovered by myself during the past season in the face of the almost vertical walls of a yawning, impassible earthquake fissure nearly opposite the mouth of Hellroaring Creek, which has evidently been visited by Indians in modern times.

INSTRUCTIONS TO WYMAN.

CAMP AT FORKS OF THE FIRE HOLE RIVER,
Yellowstone National Park, September 27, 1881.

C. H. WYMAN:

SIR: You are hereby instructed to proceed with George Rowland, and the necessary saddle, pack animals, outfit and provisions to the Lower, Midway, and Upper Geyser Basins, for the purpose of preventing vandalism of geyser cones and other objects of natural interest, and in general attend to the enforcement of the laws, rules, and regulations for the protection and management of the Yellowstone National Park.

For the prompt and full performance of this and other duties, you are hereby appointed an agent of the government, with full power of seizure of vandalized articles, and the outfits of those persons committing depredations, at your discretion, in accordance with article seven of the printed rules and regulations of the Superintendent of the Park and the Secretary of the Interior for the management thereof, published May 4, 1881, a copy of which is hereunto attached. (See appendix marked B.) You are also to use due diligence in keeping a record of the weather, making and recording observations of the periods and altitudes of the various geyser eruptions, and especially the Excelsior in the Midway Basin.

Weather permitting, you are expected to remain ten or twelve days, returning via the Norris Geyser Basin, there spending at least one day and two nights, carefully noting the geyser eruptions, and, upon reaching headquarters at the Mammoth Hot Springs, make a detailed report in writing.

P. W. NORRIS,
Superintendent Yellowstone National Park.

MAMMOTH HOT SPRINGS,
Yellowstone National Park, October 10, 1881.

P. W. NORRIS,
Superintendent of the Yellowstone National Park:

SIR: In compliance with your attached instructions of September 27, I proceeded through the Lower to the Midway Geyser Basin, carefully noting geyser eruptions, until the non-arrival of Rowland necessitated my descending the main Fire Hole River to the Marshall Hotel at night. Returned early upon the morning of the 28th, and Rowland having arrived at noon we made our camp upon the road across the Fire Hole River from the Excelsior Geyser, judging it the nearest safe place for viewing its eruptions, as well as the movements of tourists. A terribly swollen knee, from the effects of a horse kick while in the great cañon of the Gibbon, had not only thus delayed Rowland's arrival, but also, despite his earnest efforts, continued to seriously curtail his proposed observations of geyser eruptions in the Upper Basin while I was thus engaged in the Midway one. Although the attached report contains the main features of these eruptions, I may properly add that the subterranean rumblings and earth tremblings were often so fearful as to prevent sleep—so great the cloud ascending from the Excelsior Geyser, and so dense and widespread the descending spray, as to obscure the sun at mid-day, and the united mists and fogs as to saturate garments like the spray from a cataract, and often render the nights so pitchy dark as to prevent accurate observations.

Most of the rocks, hurled hundreds of feet above the column of water, fall in the foaming pond, but many are strewn over surrounding acres. This monster geyser now seems settling down to regular business, with less powerful but more frequent eruptions than during the summer, but its eruptions fully double the volume of water in the Fire Hole River, here nearly 100 yards wide, 2 or 3 feet deep, here very rapid, rendering it too hot to ford for a long distance.

Owing to Rowland's lameness, and the dense fogs in the valleys, the eruptions of the adjacent geysers, as well as those of the Lower Basin and the Geyser Meadows, were not properly noted; and, although no concert of eruptions was observable, all were unusually active and powerful. Thus also, in the Upper Basin, as noted in the occasional visits of Rowland, as well as during our two days' continuous observations there. While Old Faithful was fully sustaining her proverbial reputation for reliability, the Grand, Beehive, Castle, Splendid, and others geysers, seemed struggling to rival it; in fact, all the evidence indicates greater power and activity than during my first visit in 1875, or at any intervening period.

The recent severe snow storms tend alike to clear the park of the tourists now in it, and restrict the number of future arrivals this fall, as well as the danger of forest fires and vandalism.

En route to the Norris Geyser Basin we had a distant view of geysers in eruption in the Monument Basin, nearly amid the clouds, and others in the cañon of the Gibbon, and the Paint Pots, the appearance of all of which, as well as in the Norris Basin, indicates unusual activity. In fact, there seems no room to question the marked increase of power and activity of the internal forces throughout the Fire Hole regions.

Most respectfully, yours,

C. H. WYMAN.

Record of the eruptions of the Excelsior Geyser in the Midway Basin, Yellowstone National Park.

Date.	Time of eruption.	Duration of eruption in minutes.	Height of the column of water in feet.	Remarks.
1880.				
Sept. 27	8. 00 a. m	5	100	Witnessed the last eruption from a distance.
27	3. 30 p. m	7	75	
27	5. 30 p. m	7	100	
27	7. 15 p. m	6	90	
28	9. 00 a. m	5	60	Heavy fog in the morning, clear until sunset, and thence dense mists from the Excelsior Geyser, and fogs from the foaming, hot Fire Hole River.
28	10. 30 a. m	7	75	
28	11. 48 a. m	7	75	
28	3. 00 p. m	5	100	
28	5. 20 p. m	6	100	
28	7. 30 p. m	7	125	
29	9. 30 a. m	7	60	
29	3. 30 p. m	5	60	Heavy snow squalls, shutting off all observation after 7.20 p. m.
29	5. 00 p. m	5	70	
29	7. 20 p. m.	7	75	
30	9. 00 a. m	5	50	
30	3. 00 p. m	7	100	Heavy clouds and mists much of the day.
30	5. 20 p. m	5	125	
30	7. 15 p. m	5	75	Mists too dense for observation at night.
30	9. 30 p. m	6	75	
Oct. 1	6. 15 a. m	5	60	
1	8. 06 a. m	10	150	Cloudy and nearly dark all day.
1	10. 10 a. m	15	100	
1	12. 55 p. m	10	200	
1	3. 50 p. m	10	250	
1	5. 40 p. m	10	225	
1	7. 10 p. m	5	75	Too dense fogs and mists to continue observations.
1	9. 00 p. m	5	75	
2	12. 15 a. m	5	75	
2	3. 30 a. m	5	75	Clear, but a very heavy wind down the valley, allowing approach upon the windward side, disclosing the fact that heavy masses of the horizontally-banded wall-rock were fractured and falling into the foaming cauldron, which was all that could be observed, save an occasional rock eruption.
2	6. 45 a. m	5	75	
2	8. 15 a. m	5	75	
2	10. 10 a. m	5	75	
	12. 15 p. m	4	60	

*Record of the eruptions of the Excelsior Geyser in the Midway Basin, &c.—*Continued.

Date.	Time of eruption.	Duration of eruption in minutes.	Height of the column of water in feet.	Remarks.
1880. Oct. 2	2. 15 p. m	5	50	
2	4. 15 p. m	7	200	
2	5. 30 p. m	5	75	
2	7. 00 p. m	5	50	
2	9. 05 p. m	5	50	
2	11. 15 p. m	5	60	
3	6. 30 a. m	5	100	
3	8. 00 a. m	10	150	Countless rocks, of many pounds weight, hurled like a rocket high above the column of water, some of which fell in and across the river, which is here 100 yards wide, and during much of the day was a foaming flood of hot water.
3	10. 10 a. m	10	300	
3	12. 30 p. m	10	75	
3	3. 00 p. m	10	250	
3	4. 30 p. m	7	75	
3	5. 45 p. m	5	80	
3	7. 25 p. m	6	75	
3	9. 20 p. m	5	75	
3	11. 30 p. m	5	75	
4	6. 00 a. m	5	75	
4	7. 30 a. m	5	75	Broke camp and went to the Upper Basin at 9 a. m.
4	9. 00 a. m	7	75	
4	10. 20 a. m	10	150	
4	11. 45 a. m	5	150	
6	3. 00 p. m	5	75	Returned through mist and snow squalls; weather quite cold.
6	5. 25 p. m	7	100	
6	7. 19 p. m	6	80	
6	9. 00 p. m	7	120	
6	10. 40 p. m	5	75	
7	3. 45 a. m	6	80	Clear and cold, but dense fogs along the river for miles.
7	5. 20 a. m	7	125	
7	6. 45 a. m	5	100	
7	9. 08 a. m	7	120	Left the basin for the Norris Geyser.

ERUPTIONS OF GEYSERS IN THE UPPER BASIN.

Old Faithful—This typical geyser during our visit seemed to be in greatest activity and power, having hourly eruptions of five minutes' duration, and columns of water 175 feet high.

GRAND.

Date.	Time of eruption.	Duration of eruption in minutes.	Height of the column of water in feet.	Remarks.
1881. Oct. 4	9.45 a. m	20	200	Observed by Rowland. The column of water at all of these eruptions was vertical and of remarkable symmetry and beauty.
4	5.10 p. m	25	200	
5	3.25 p. m	20	200	
6	9.15 a. m	20	200	
6	4.20 p. m	20	200	

SPLENDID.

Date.	Time of eruption.	Duration of eruption in minutes.	Height of the column of water in feet.	Remarks.
1880. Oct. 4	7.15 a. m			Eruptions uniformly much like those of Old
4	9 a. m..................			Faithful, but the form of the column of water
4	11 a. m.................			less vertical and more spreading.
4	2.30 p. m			
4	6.30 p. m			
5	6 a. m.................			
5	8.20 a. m			
5	11.20 a. m			
5	1.15 p. m			
5	3.45 p. m			
5	6.30 p. m			
6	6 a. m.................			
6	8.30 a. m			
6	11 a. m.................			
6	1.20 p. m			

CASTLE.

Date.	Time of eruption.	Duration	Height	Remarks.
Oct. 4	3 p. m..................	25	75	There was a constant agitation and several small
6	9.45 a. m..............	30	100	eruptions.

BEEHIVE.

Date.	Time of eruption.	Duration	Height	Remarks.
Oct. 4	9.45 p. m	5	175	Column of water always vertical, and of great
5	2.15 p. m	5	200	symmetry and beauty.
6	8.40 p. m	5	180	

GIANT.

Date.	Time of eruption.	Duration	Height	Remarks.
Oct. 5	8 p. m.................	25	250	The accompanying earth-trembling was terrific.

The Lion, Lioness, Grotto, Fan, Riverside, Saw-mill, and other geysers had eruptions during the night, which we failed to properly observe, but, from the noise of their spouting, all were in full force and activity.

LOWER GEYSER BASIN.

Fountain.—Usually had an eruption each forenoon, those observed being of from 10 to 15 minutes' duration, with water column from 60 to 90 feet high, and very spreading. Rowland's lameness and the dense fogs prevented extended observations in the Lower Basin, as well as in the Geyser Meadows.

NORRIS GEYSER BASIN.

Monarch.

Date.	Time of eruption.	Duration of eruption in minutes.	Height of the column of water in feet.	Remarks.
Oct. 8	6.20 a. m	20	100	The eruptions are simultaneously through three
9	6.30 a. m	25	125	orifices—2 by 12, 2½ by 11, and 5 by 6 feet, respectively, their combined flow producing for the time a large sized stream of hot water.

New Crater.—Exhibits two kinds of eruptions—one of them, each half hour, 50 feet high, and another about 100 feet high daily.

Minute Man.—Eruptions 25 or 30 feet high each minute, with little variation.

Emerald.—Evidently has an occasional eruption, although none were observed.

Vixen.—Eruption from 40 to 50 feet high, each two or three hours.

Constant, Twins, Triplets, and many others in the Porcelain Vale, seem in nearly constant eruption, so that the spray and fogs greatly obscure the sun's rays by day, and render the nights dark, damp, and unpleasant.

Report of weather in the Geyser Basins.

MIDWAY BASIN.

Date.	Thermometer.			Remarks.
	Sunrise.	Noon.	Sunset.	
1881.				
Sept. 27 ..	32	50	38	Cloudy.
28 ..	38	49	42	Clear; heavy mist from the Excelsior Geyser.
29 ..	40	55	32	Snow-squalls.
30 ..	26	· 52	30	Heavy clouds and mist.
Oct. 1 ..	36	50	32	Do.
2 ..	32	60	44	Clear, but windy.
3 ..	34	61	40	Clear, but windy; dense mist at night.
4 ..	26	....	....	

UPPER BASIN.

Oct. 4 ..	....	64	46	Dense mists from geysers.
5 ..	25	68	42	Clear morning; thunder-shower at 2 p. m.
6 ..	32	38	....	Snow-squalls and blinding mists.

MIDWAY BASIN.

Oct. 6 ..	....	....	30	Snow-squalls and blinding mists.
7 ..	33	....	....	Clear, but very windy. Went to the Norris Geyser Basin.

NORRIS BASIN.

Oct. 7 ..	....	....	32	
8 ..	18	50	40	Clear and lovely.
9 ..	16	....	....	Clear day. Left for headquarters at 7.20 a. m., arrived at 12 m.

GEYSERS.

The theories regarding these and other kinds of hot springs in the park were so fully treated of in my report of last year, and the records of their eruption, notably during the latter part of this season, in the foregoing trustworthy report of Wyman, leaves but litte necessary to show that, with the exception of the local changes at the Mammoth Hot Springs and of the Safety-Valve Basin in the Grand Cañon, there is evidently a far greater development of power than ever before witnessed throughout the entire Fire-Hole regions. But as to the cause or causes, probable duration, or future tendencies, we only know that they are at variance with the accepted and apparently correct theory of their dwindling character, with one marked exception. This is in the Midway Basin of the Fire Hole River, where the evidence is conclusive of not only spasmodic, but continuous increase of power.

The following description is from Hayden's Report of 1871, pp. 114, 115:

About three miles up the Fire-Hole from Camp Reunion we meet with a small but quite interesting group of springs on both sides of the stream. There is a vast accumulation of silica, forming a hill 50 feet above the level of the river. Upon the summit is one of the largest springs yet seen, nearly circular 150 feet in diameter; boils up in the center, but overflows with such uniformity on all sides as to admit of the formation of no real rim, but forming a succession of little ornamental steps, from 1 to 3 inches in height, just as water would congeal from cold in flowing down a gentle declivity. There was the same transparent clearness, the same brilliancy of coloring to the waters; but the hot steam and the thinness of the rim prevented me from approaching it near enough to ascertain its temperature or observe its depth, except at one edge, where it was 180°. It is certainly one of the grandest hot springs ever seen by human eye. But

FIG. 27.—Excelsior Geyser, 1872.

the most formidable one of all is near the margin of the river. It seems to have broken out close by the river, to have continually enlarged its orifice by the breaking down of its sides. It evidently commenced on the east side, and the continual wear of the under side of the crust on the west side has caused the margin to fall in, until an aperture at least 250 feet in diameter has been formed, with walls or sides 20 to 30 feet high, showing the laminæ of deposition perfectly. The water is intensely agitated all the time, boiling like a caldron, from which a vast column of steam is ever arising, filling the orifice. As the passing breeze sweeps it away for a moment, one looks down into this terrible, seething pit with terror. All around the sides are large masses of the siliceous crust that have fallen from the rim. An immense column of water flows out of this caldron into the river. As it pours over the marginal slope, it descends by numerous small channels, with a large number of smaller ones spreading over a broad surface, and the marvelous beauty of the strikingly vivid coloring far surpasses anything of the kind we have seen in this land of wondrous beauty; every possible shade of color, from the vivid scarlet to a bright rose, and every shade of yellow to delicate cream, mingled with vivid green from minute vegetation. Some of the channels were lined with a very fine, delicate, yellow, silky material, which vibrates at every movement of the waters. Mr. Thomas Moran, the distinguished artist, obtained sketches of these beautiful springs, and from his well-known reputation as a colorist, we look for a painting that will convey some conception to the mind of the exquisite variety of colors around this spring. There was one most beautiful funnel-shaped spring, 20 feet in diameter at the top, but tapering down, lined inside and outside with the most delicate decorations. Indeed, to one looking down into its clear depths, it seemed like a fairy palace. The same jelly like substance or pulp to which I have before alluded covers a large area with the various shades of light red and green. The surface yields to the tread like a cushion. It is about 2 inches in thickness, and, although seldom so tenacious as to hold together, yet it may be taken up in quite large masses, and when it becomes dry it is blown about by the wind like fragments of variegated lichens.

The above, cut from the Hayden report of 1872, and the description thereof in that of 1871, are here republished, both for their accuracy and as a datum from which to trace subsequent and future developments. This clearly proves the comparatively recent outburst of the yawning pool of hot water, in border parlance heretofore called "Hell's Half Acre," which during the past season has fully justified the name and greatly exceeded the dimensions. Although noted for the deep ultra-marine blue, ever-agitated waters, so characteristic of the true geyser when not in eruption, there was neither evidence nor indications of recent eruptions until late in August, 1878. I then distinctly heard its spoutings when near Old Faithful, 6 miles distant, but arrived too late to witness them, though not its effects upon the Fire Hole River, which was so swollen as to float out some of our bridges over rivulet branches below it.

Crossing the river above the geyser and hitching my horse, with bewildering astonishment I beheld the outlet at least tripled in size, and a furious torrent of hot water escaping from the pool, which was shrouded in steam, greatly hiding its spasmodic foamings. The pool was considerably enlarged, its immediate borders swept entirely clear of all movable fragments of rock, enough of which had been hurled or forced back to form a ridge from knee to breast high at a distance of from 20 to 50 feet from the ragged edge of the yawning chasm. Perhaps no published statement of mine in reference to the Wonder Land has ever more severely tested the credulity of friends or of the public; and even General Crook and Secretary Schurz, to whom I pointed out the decreasing proofs of this eruption, seemed to receive it with annoying evidences of distrust. The volume of steam arising from this pool continued to increase until, on reaching the Lookout lower border of the valley, late in November, 1880, it appeared so great as to cause me to visit it the next day, hopeful of seeing an eruption or evidences of a recent one. This I failed to find, but not a volume of steam which then shrouded all near it, as it did the whole of the lower valley before the next morning. In order to make the Mammoth Hot Springs, 40 miles distant, that day, I started early, and with the thermometer but little above zero groped my way through this fog, which chilled to the marrow, to the Lookout Terrace, 3 miles from the Forks of the Fire Holes and 8 from the geyser, and emerged therefrom by ascending above it into a broad and brilliant scene of beauty seldom witnessed by human vision. From the foaming half-acre caldron an enormous column of steam and vapor constantly arose, at first vertically, then swayed by a moderate but steady southern wind northerly, increasing with the altitude, until intermingling with or forming a cloud at the proper elevation, from which a nearly imperceptible descending vapor, carried northerly, covered and loaded to pendency the southern branches of the dark pine and fir fringes to the terrace slopes and craggy cliffs of the Madison Plateau, to its great cañon beyond the Gibbon, fully 15 miles from this earthly Gehenna.

FIG. 28.—Excelsior Geyser, 1881.

Beneath this unique cloud-awning the low and seemingly distant rays of a cold, cloudless sunrising, in struggling through this vapor-laden atmosphere, formed a variety of tints and reflections from the inimitably beautiful festoons of frost formation, while commingled with a dark green background of foliage, of somber cliffs and snowy mountains—a brilliancy of blended wavy shades and halos enchantingly beautiful. This was my parting view of that geyser last year; and before my return this season, great changes had occurred. From the statements of G. W. Marshall, at the Forks of the Fire Holes, February was ushered in by dense fogs and fearful rumblings and earth tremblings, which he ultimately traced to regular eruptions, daily, or rather nightly, commencing about 10 o'clock p. m., gradually recurring later, until by July 1 they were after daylight; and this eruption is now about 10 a. m., showing a loss of twelve hours in nine months. During much of the summer this eruption was simply incredible, elevating to heights of from 100 to 300 feet sufficient water to render the rapid Fire Hole River, nearly 100 yards wide, a foaming torrent of steaming hot water, and hurling rocks of from 1 to 100 pounds weight, like those from an exploded mine, over surrounding acres. By far the finest landmark that I ever beheld in all my mountain wanderings was the immense column of steam, even when the geyser was not in eruption, always arising from this monster, which was ever plainly visible to where, at the proper elevation, it formed a cloud that floated away in a long line to the leeward in the clearest summer's day, and was never to be mistaken for any other wherever seen, which was upon all the surrounding mountains, including the Rocky and Shoshone ranges, portions of which that I visited were fully 100 miles distant. In September the eruptions branched into one about 4 o'clock p. m., and soon after to others, until it now seems to be settling down to regular business as a two or three hour geyser, so immeasurably excelling any other, ancient or modern, known to history, that I find but one name fitting, and herein christen it the "Excelsior" until scientists, if able, shall invent a more appropriate one. This pool is now 400 paces in circumference.

The Fire Hole River is down a declivity of some 20 or 30 feet from where the outlet beside the horseman is shown in the Hayden view (Fig. 27), Wyman's camp being across the river, still eastward—and many rocks were hurled into or across it, and also to the great spring, with the steam cloud in the background, as well as another, sixty paces to the north of the geyser, whose brilliantly colored outlet is shown as joining that of the geyser upon the brink of the declivity to the river, in the above view from my sketch (Fig. 28), which was taken at a period of less activity between the regular daily eruptions early in the season than observed at any subsequent period.

REPORT OF GAMEKEEPER.

MAMMOTH HOT SPRINGS,
YELLOWSTONE NATIONAL PARK,
September 30, 1881.

SIR: I hereby respectfully submit the following report of my operations as gamekeeper of the park, for the protection of its animals, since furnishing my report of November 25, 1880, from the gamekeeper's cabin, near the confluence of the Soda Butte and the East Fork of the Yellowstone River. I there remained, sometimes having George Rowland or Adam Miller for a comrade, but often alone, during the entire winter, the early part of which was so severe that there were no mountain hunters—the Clarke's Fork miners twenty miles distant one way, and the boys at the headquarters nearly forty the other, being the nearest, and in fact the only men in these regions. The snowfall was unusually great, and remained very deep high in the mountains, but the winds and hot vapors from the Fire Hole Basin at the foot of Mount Norris kept the snow pretty clear along its western slopes, where there were abundance of mountain sheep, and some elk, all winter. Elk to the number of about 400 wintered in small bands in the valleys of the East Fork and Soda Butte, where the snow was about knee-deep. The Slough Creek and Hellroaring bands of bison did not venture near the cabin until February, nor did those of Amethyst Mountain at all; and the most of the deer and antelope descended into the lower Yellowstone Valley early in the winter.

The most of the Clarke's Fork miners seemed disposed to kill only what game they needed for food, and preserve the rest from slaughter for their hides only, and hence I returned to the headquarters in the spring, which opened very early and continued warm and pleasant. This allowed me to visit many other portions of the park, some-times on snow-shoes and sometimes with saddle and pack-horses. I found that very few of the deer or antelope wintered anywhere in the park; that a small band of bison wintered on Alum Creek, and another on the South Fork of the Madison; that there were elk in nearly all of the warm valleys, and moose around the Shoshone and the fingers of the Yellowstone lakes; big-horn sheep on all the mountain slopes; wolverine, marten, and various kinds of foxes, who do not leave the park in winter, nor do the bears of all kinds, as they hibernate. During the remainder of the season I have been active in the various duties of killing what game was necessary for our various parties of laborers, and protecting the rest from wanton slaughter by some of the tourists and a band of Bannock Indians on the North Madison. I also guided the party of Governor Hoyt and Colonel Mason from the Two Ocean Pass to the Fire Holes, and accompa-nied you in the long and arduous exploration of the Sierra Shoshone, and the Rocky Mountain, from Turbid Lake to Mount Sheridan; and in a final tour of the main roads and trails of the park close my services and resign my position as gamekeeper of the park to resume private enterprises now requiring my personal attention. The unfor-tunate breakage of my thermometer when it could not be replaced prevented my keep-ing other than a record of fair and stormy days, winds and rain and snow-fall during last winter, a synopsis of which is hereunto attached.

In conclusion, I may justly add that my relations with yourself, with your men, and with nearly all of the visitors to the park, as well as the surrounding miners and hunters have always been most cordial; but, as stated in my report of last year, I do not think that any one man appointed by the honorable Secretary, and specifically designated as a gamekeeper, is what is needed or can prove effective for certain necessary pur-poses, but a small and reliable police force of men, employed when needed, during good behavior, and dischargeable for cause by the superintendent of the park, is what is really the most practicable way of seeing that the game is protected from wan-ton slaughter, the forests from careless use of fire, and the enforcement of all the other laws, rules, and regulations for the protection and improvement of the park.

Most respectfully, yours,

HARRY YOUNT,
Gamekeeper.

P. W. NORRIS,
Superintendent of the Yellowstone National Park.

OBSERVATIONS OF WEATHER.

November.—From the 26th to the 30th, inclusive, snowy.
December.—During this month, one day was rainy, two hazy, six clear, cold, and windy, and twenty-two snowy.
January.—The 13th, 16th, 17th, 18th, 20th, 21st, 22d, 24th, and 25th, nine days, were clear; the remainder of the month snowy, and mainly very cold.
February.—The 2d and 3d, two days, rainy; 14th, one day, was clear; the 8th, snowy; the 9th, squally, and twenty-three days snowy.
March.—Twenty-four days were clear, and mostly mild, and some warm; one day rainy, two snowy, and four cloudy.
April.—The 1st, 4th, 5th and 7th were clear, the 2d, 3d, and 6th rainy, and the snow so soft that traveling with my Norwegian snow shoes 14 feet long, was hard work, and leaving them at the middle fall of the Gardiner, went thence through the cañon to the boys at headquarters, they keeping the weather records correctly there-after.

INTRODUCTION TO ROADS, BRIDLE-PATHS, AND TRAILS.

In preceding reports I have followed the usual custom of calling all traveled routes either roads or trails, but it having become, as it will con-tinue, necessary to mention mountain, fire-hole, cliff, and cañon trails for footmen only, as well as those in common use for saddle and pack ani-mals, the latter are herein tabled as bridle-paths, the former as trails; while the lodge-pole or Indian and game trails only are thus designated whenever mentioned in the body of this report. I have, also, in some of my preceding reports, stated that, as none of our roads, bridle-paths, or trails had ever been measured, the tables of them were at best only approxi-mations, and the distances therein shown are more probably over than under estimated. This view the odometer measurements of Capt. W. S.

Stanton, Corps of Engineers, and of First Lieut. E. Z. Steever, Third Cavalry, made during July and August of the past season, have proven correct, and it is one of the amusing incidents in connection with these peculiar regions that while prominent judges, senators, governors, and other officers of the government were making me the subject of their raillery upon the annoying length of my estimated miles, other officers were by actual measure proving many of them far too short. This is especially noticeable in the direct or Mammoth Hot Spring road, estimated when made, in 1878, as 50 miles in length, and which was nearly correct at that time, but it having been materially shortened by changing the road from the cañon to the plateau of the Madison, a cut-off through the earthquake region and somewhat elsewhere, it is now found to be less than 37 miles long, which is only about one-half of the Mount Wabash route, and can never be essentially shortened. The tables of distances, as received from Captain Stanton and Lieutenant Steever, were well arranged and computed, evincing accurate odometer measurements, and are accepted and used as such; but owing to the subsequent construction of new roads and bridle-paths, or changes in old ones, as well as from their want of knowledge of the names of many places which it is believed essential to mention, these tables are thus amended; but all portions of them have been accepted which were proper to use, and are credited and indicated by a *.

SYNOPSIS OF ROADS, BRIDLE-PATHS, AND TRAILS IN THE YELLOWSTONE NATIONAL PARK.

	Between points.	Total.
Road towards Bozeman.	Miles.	Miles.
* From headquarters at the Mammoth Hot Springs to northern boundary line of Wyoming	....	1.99
Northern boundary line of the National Park, below the mouth of the Gardiner River	5.00	6.99
Direct road to the Forks of the Fire-Hole River.		
* From headquarters at the Mammoth Hot Springs to Terrace Pass	....	1.93
* Swan Lake	3.21	5.14
* Crossing of Middle Fork of Gardiner River	2.33	7.47
Willow Park, upper end	3.50	10.97
* Obsidian Cliffs and Beaver Lake	1.37	12.34
* Green Creek	1.40	13.74
* Lake of the Woods	.76	14.50
* Hot Springs	1.68	16.18
* Norris Fork Crossing	4.17	20.35
* Norris Geyser Basin	.71	21.06
* Geyser Creek and Forks of the Paint-Pot trail	3.13	24.19
* Head of Cañon of the Gibbon and foot-bridge on trail to Monument Geysers	.72	24.91
* Falls of the Gibbon River	3.75	28.66
* Cañon Creek	.59	29.25
Earthquake Cliffs	3.00	32.25
* Lookout Terrace	1.50	33.75
* Marshall's Hotel, at the Forks of the Fire Hole River	2.43	36.18
Road from Forks of the Five Hole River to foot of the Yellowstone Lake.		
From Marshall's Hotel to forks of the road near Prospect Point	....	1.00
* Hot Springs	1.08	2.08
* Rock Fork	3.86	5.94
Willow Creek	2.00	7.94
Foot of the grade up the Madison Divide	2.00	9.94
Upper end of Mary's Lake	1.91	11.85
* Sulphur Lakes and Hot Springs	1.12	12.97
Alum Creek Camp	2.00	14.97
Sage Creek Crossing	2.00	16.97
Fork of the road to the falls near the Yellowstone River	5.00	21.97
Mud Geysers	2.00	23.97
Grizzly Creek	3.00	26.97
* Foot of the Yellowstone Lake	3.26	30.23

Roads, bridle-paths, and trails in the Yellowstone National Park—Continued.

	Between points.	Total.
Branch road to the Great Falls of the Yellowstone.		
	Miles.	*Miles.*
From Forks of the Fire Hole River to forks of the lake road to the Great Falls, as above	.. ..	21. 97
Sulphur Mountain	1. 50	23. 47
* Alum Creek	1. 61	25. 08
* Upper Falls of the Yellowstone, bridle-path	3. 26	28. 34
* Crystal Falls and Grotto Pool, bridle-path	. 40	28. 74
* Lower (Great) Falls of the Yellowstone	. 24	28. 98
Road to Tower Falls.		
* Headquarters at the Mammoth Hot Springs to bridge over the Gardiner River	....	1. 77
* Bridge over the East Fork of the Gardiner River	. 38	2. 15
* Upper Falls to East Fork of the Gardiner River	2. 06	4. 21
* Black Tail Deer Creek	2. 70	6. 91
Lava Beds	2. 00	8. 91
* Dry Cañon, or Devil's Cut	4. 69	13. 60
* Pleasant Valley	2. 28	15. 88
* Forks of the Yellowstone	2. 48	18. 36
* Tower Falls	3. 19	21. 55
Geyser Basin road.		
* Marshall's Hotel to forks of road at Prospect Point	....	1. 00
* Old Camp Reunion	1. 00	2. 00
Fountain Geyser in the Lower Geyser Basin	1. 00	3. 00
* Excelsior Geyser, in the Midway Geyser Basin	2. 00	5. 00
* Old Faithful, in the Upper Geyser Basin	6. 00	11. 00
Madison Plateau road.		
Marshall's Hotel to Forest Spring	....	3. 00
* Marshall's Park	2. 12	5. 12
* Lookout Cliffs	3. 59	8. 71
Riverside Station and Forks of Kirkwood or Lower Madison Cañon road to Virginia City	3. 52	12. 23
Bridge over South Madison River	11. 53	23. 76
Madison Cañon road.		
Marshall's Hotel to forks of road to the Mammoth Hot Springs	....	4. 00
Mouth of the Gibbon River	5. 00	9. 00
Foot of the Madison Cañon	6. 00	15. 00
Riverside Station	3. 00	18. 00
Queen's Laundry road.		
Marshall's Hotel to crossing Laundry Creek	....	1. 00
Twin Mounds	1. 00	2. 00
Queen's laundry and bath-house	. 50	2. 50
A bridle-path 3 miles long extends from there to the Madison Plateau road, and another is partially completed *via* Twin Buttes and Fairy Falls to the Midway Geyser Basin.		
Middle Fork of the Gardiner bridle-path.		
Headquarters at the Mammoth Hot Springs to the West Gardiner	....	2. 00
Falls of the Middle Gardiner	2. 00	4. 00
Sheepeater Cliffs	2. 00	6. 00
Road to the Geysers	1. 00	7. 00
Painted Cliff bridle-path.		
Meadow Camp to head of Grand Cañon	....	1. 00
Safety Valve Pulsating Geyser	1. 00	2. 00
Yellowstone River at Painted Cliffs	1. 00	3. 00
Paint Pots bridle-path.		
Mouth of Geyser Creek to the Paint Pots	....	1. 00
Geyser Gorge	1. 00	2. 00
Earthquake Gorge	2 00	4. 09
Rocky Fork Crossing	2. 00	6. 00
Mary's Lake Road, near Yellowstone Creek	5. 00	11. 00
Mount Washburn bridle-path.		
* Tower Falls to Forks of Trail	....	1. 87
* To Summit of Mount Washburn	4. 13	6. 00
Cascade Creek	7. 22	13. 22
* Great Falls of the Yellowstone	2. 00	15. 22

5 Y P

Roads, bridle-paths, and trails in the Yellowstone National Park—Continued.

	Between points.	Total.
Grand Cañon bridle-path.		
	Miles.	Miles.
* Tower Falls to Forks of Trail		1. 87
Antelope Creek	4. 00	5. 87
Rowland's Pass of Mount Washburn	2. 00	7. 87
Glade Creek	2. 47	10. 34
* Mud Geyser	1. 00	11. 34
* Hot Sulphur Springs	. 83	12. 17
* Meadow Camp and fork of Painted Cliffs bridle-path Trail	1. 59	13. 76
Brink of the Grand Cañon	1. 00	14. 76
* Lookout, Paint, and forks of trail into the cañon below the falls	2. 19	16. 95
* Great Falls of the Yellowstone	. 74	17. 69
Shoshone Lake bridle-path.		
* Old Faithful, in the Upper Geyser Basin, to Kepler's Cascades		1. 94
* Leech Lake	2. 72	4. 66
Norris Pass, Continental Divide	3. 00	7. 66
DeLacey Creek, Pacific waters	. 97	8. 63
* Two-Ocean Pond, on Continental Divide	3. 50	12. 13
* Hot Springs, at head of thumb of the Yellowstone Lake	2. 99	15. 12
* Hot Spring, on Lake Shore	2. 02	17. 14
* Hot Spring Creek	4. 00	21. 14
* Natural Bridge	7. 44	28. 58
* Outlet of Yellowstone Lake	4. 68	33. 26
Miners' bridle-path.		
* Baronette's Bridge, at forks of the Yellowstone River, to Duck Lake		1. 76
* Amethyst Creek	8. 30	10. 06
* Crossing, East Fork of Yellowstone River	2. 16	12. 22
Gamekeeper's Cabin	. 50	12. 72
* Soda Butte, medicinal springs	2. 65	15. 37
Trout Lake	2. 00	17. 37
* Round Prairie	3. 00	20. 37
North line of Wyoming	3. 84	24. 21
* Clarke's Forks Run Camp, near northeast corner of the park	3. 18	27. 39
Hoodoo or Goblin Mountain bridle-path.		
Gamekeeper's cabin, on the Soda Butte, to Hot Sulphur Springs		2
Ford of Cache Creek	1	3
Alum Springs and return	4	7
Calfee Creek	4	11
Miller's Creek	2	13
Mountain Terrace	8	21
Old Camp	5	26
Goblin Labyrinths	2	28
Monument on Hoodoo Mountain	1	29
Fossil Forest bridle-path.		
Summit of Amethyst Mountain		3
Gamekeeper's cabin to foot of Mountain	3	6
Orange Creek	5	11
Sulphur Hills	4	15
Forks of Pelican Creek	8	23
Indian Pond at Concretion Cove of the Yellowstone Lake	5	28
Lower Ford of Pelican Creek	3	31
Foot of the Yellowstone Lake	3	34
Passamaria or Stinkingwater bridle-path.		
Concretion Cove to Turbid Lake		3
Jones' Pass of the Sierra Shoshone Range	7	10
Confluence of the Jones and Stinkingwater Fork of the Passamaria River	12	22
Nez Percé bridle-path.		
Indian Pond to Pelican Valley		3
Ford of Pelican Creek	3	6
Nez Percé Ford of the Yellowstone	6	12
Alum Creek bridle-path.		
From the Great Falls of the Yellowstone, via Crystal Falls and Grotto Pool and the Upper Falls, to the mouth of Alum Creek	4	4

Roads, bridle-paths, and trails in the Yellowstone National Park—Continued.

	Between points.	Total.
Terrace Mountain Trail.	*Miles.*	*Miles.,*
Headquarters at the Mammoth Hot Springs, amongst the numerous active and extinct Mammoth Springs, to foot of the Ancient Terraces		1
Up steep pine, fir, and cedar clad terraces, to summit of the mountain	1	2
Along the range of the vertical cliffs, for 400 to 800 feet high	2	4
Descent of South Terrace to Rustic Falls, 40 feet high, at the head of the impassable cañon of the West Fork of the Gardiner River	1	5
Upon the southern cliff, above these falls, is a Sheepeater arrow-covert, and the remains of an ancient game-driveway thereto.		
Swan Lake, on the Fire Hole road	1	6
Trail to the Falls of he East Gardiner River.		
From the road near the middle of the cañon along the eastern declivity, one mile... To the fall, not unlike the famous Minnehaha, and like which, allows a safe pathway between the sheet of water and the wall rock.		1
Monument Geyser Trail.		
Foot-bridge at head of the cañon of the Gibbon, which ascends nearly 1,000 feet within a distance of one mile, some portions of which are exceedingly difficult for a horseman, and hence called a trail. The active and the extinct and crumbling geyser cones are alike uniquely interesting, and the outlook remarkably beautiful.		1

Trail, or footpath, to head of the Great Falls of the Yellowstone.

Leaves at the lower end of the camping ground above, and descends 500 or 600 feet within one-fourth of a mile to the pole-bordered outlook at the very head of the cataract.

Trail to the Yellowstone River below the Lower Falls of the Yellowstone.

This trail descends Spring Run from the rustic bridge nearly to its waterfall, thence along the steep declivity beneath Lookout Point, in a winding, dangerous way, to the foaming river, which cannot now be ascended, along it, as formerly, to the foot of the falls upon this side; but can be reached upon the other, via the timber-fringed gorge.

The main danger is from detached fragments of rock, which attain incredible velocity before reaching the river.

Besides these trails there are several others to fossil forests, cliffs, geyser or sulphur basins or falls, which will be fully noted in the forthcoming guide-book of the Park.

RECAPITULATION OF DISTANCES, ROADS, BRIDLE-PATHS, AND TRAILS WITHIN THE PARK.

ROADS.

	Miles.
1. Road to the north line of the Park, towards Bozeman, about	7.00
2. Direct road to the Forks of the Fire Hole Rivers	36.00
3. Road from Forks of the Fire Hole Rivers to the foot of the Yellowstone Lake, about	30.00
4. Branch road from Sage Creek to Alum Creek	4.00
5. Tower Falls road, about	21.50
6. Geyser Basin road	11.00
7. Madison Plateau	24.00
8. Madison Cañon	18.00
9. Queen's Laundry	2.50
	153.00

BRIDLE-PATHS.

1. Middle Gardiner	7.00
2. Painted Cliffs	3.00
3. Paint Pots	11.00
4. Mount Washburn	15.00
5. Grand Cañon from the Forks, about	16.00
6. Shoshone Lake	33.00
7. Mines, to Clark's Fork, about	27.00
8. Hoodoo or Goblin Mountain	29.00
9. Fossil Forest	34.00

	Miles.
10. Passamaria	22. 00
11. Nez Percé Ford	12. 00
12. Alum Creek	4. 00
	213. 00

TRAILS.

1. Terrace Mountain	6. 00
2. Falls of the East Gardiner	1. 00
3. Monument Geyser	1. 00
4. To head of Great Falls of the Yellowstoue, about 200 yards.	
5. To river below the Great Falls of the Yellowstone, 200 yards.	
	8. 00

RAILROADS.

Two railroads have entered Montana, the Northern Pacific being now completed to the vicinity of Miles City, at the mouth of Tongue River, upon the Yellowstone, about 300 miles below its Gate of the Mountains, which they promise to reach during 1882, and soon thereafter run a branch up the tolerably smooth open valley of the Yellowstone to the mouth of the Gardiner, ascending it to the great Hot Medicinal Spring, where application has been made by desirable parties for the establishment of a sanitarium, one mile below the Mammoth Hot Springs and about sixty miles from their main line. The Utah Northern Railroad is completed from Ogden to Silver Bow, near Butte, and is now engaged in surveying the route of a branch by way of Ruby Valley, Virginia City, and the Upper Madison, to the Forks of the Fire Holes, a distance of about 140 or 150 miles from the main line at Dillon. With little doubt, one or both of these roads will enter the Park within two or three years hereafter, and ultimately a connection by the latter, through the valleys and cañoned branches of the Madison and the Gallatin, skirt the western border of the Park from the Forks of the Fire Holes to Bozeman, on the line of the Northern Pacific Railroad.

Should the mining developments of these mountain regions equal present indications, a railroad will reach the Park from the east via Clarke's Forks Mines or the Two Ocean Pass, or both of them, within a few years hereafter. The approach of these railroads—notably the Utah Northern—materially facilitates reaching the Park, which each road as they near it, will increase accessibility, and will soon invite a healthy competition for the patronage of tourists in making a cheap, rapid, and easy visit to the Wonder Land; planning it as the turning point, as well as the main region of attraction, in a season's ramble for health and enjoyment.

Should these anticipations be realized a visit to the Park will become national in character and popular with our people, so that ere long the flush of shame will tinge the cheeks of Americans who are obliged to acknowledge that they loiter along the antiquated paths to pigmy haunts of other lands, before seeking health, pleasure, and the soul expanding delights of a season's ramble amid the peerless snow and cliff encircled marvels of their own.

There is now assurance of increased facilities for conveyance of tourists from Bozeman, nearly 80 miles through Trail Pass, and up the Yellowstone Valley to the headquarters of the Park at the Mammoth Hot Springs, and from Virginia City some 95 miles via the old Henry's Lake route, or 90 miles by the new one up the Madison to Riverside, which was constructed during the past season by Judge Kirkwood for the spirited citizens of that town, to the Forks of the Fire Hole River, and also by the practical use of the old route via Henry's Fork and Lake, which the odometer measurements of Lieutenant Steever during the past season make 103 miles from the Forks of the Fire Hole River to Beaver Cañon, and practically about the same distance to Camas Station, both upon the Utah Northern Railroad, in the Snake River Valley, below the mountains. Believing it to be a necessity, it is now my purpose to issue a guide-book of the Park, containing a map, illustrations, and descriptions of various objects of interest, routes of approach, list of articles necessary for camp outfit and provisions, approximate time, and cost of a tour of the Wonder Land, in time for the use of next season's tourists thereto.

CONDENSED SUMMARY OF THE SEASON'S EXPLORATIONS' WORK—RECOMMENDATIONS.

For the purpose of concisely showing what has been accomplished in the Park during the past season, as well as what is considered essential to be done therein during the next, the following synopsis of each is added:

SYNOPSIS OF THE PAST SEASON'S OPERATIONS.

The following explorations have been made: Nearly all of the Madison or Mary's Lake Divide, with several brimstone basins, and also passes to Violet Creek, to the Norris Fork of the Gibbon, and to the Paint Pots bordering the Gibbon Meadows, of a nearer route to the Hoodoo region, and additional Labyrinths of Goblins upon the Passamaria and elsewhere of an open lovely pass connecting the Pelican Valley with that of the East Fork of the Yellowstone. The first general exploration of the Sierra Shoshone range, or eastern border of the Park, which is known to have ever been made by white men, including a very low and direct pass from the Passamaria Cañon to the Yellowstone Lake. An examination was also made of the main Rocky Mountain portion of the southern border of the Park from the Two Ocean Pass via Phelps's Pass, and various unknown fountain heads of the Snake River branch of the Columbia, Mount Sheridan and Heart and Riddle Lakes to the Thumb of the Yellowstone, including the discovery of some fine valleys and passes.

IMPROVEMENTS MADE.

Buildings constructed.—Hopeful of a saw-mill and cheaper lumber, the only buildings constructed during the past season were:
A small, earth-covered vault or detached fire-proof store-room for the safety of much of our provision, tools, and camp outfit at our headquarters
A double-roomed earth-roofed bath house at the matchless Queen's Laundry, near the forks of the Fire Hole Rivers, together with wooden troughs for conveying water thereto, for the free use of the public. A line of wooden troughs for the purpose of conducting the Terrace-building waters to and successful recoating and building up of the extinct pulsatory Geyser Cone, called Devil's Thumb, at the Mammoth Hot Springs.
Bridges Constructed.—One amid the spray at the head of the Upper Falls of the east fork of the Gardiner River. A bridge over the main Blacktail Creek near its forks, and another over Elk Creek near the Dry Cañon. Three bridges in the valley of the East Fork of the Fire Hole, two upon Alum Creek; two upon Sage Creek and two upon Hot Spring Creek, all upon the new road to the Yellowstone Lake, and several others upon the Shoshone Bridle Path across the Continental divide

to the said lake. Also two foot bridges across the Fire Hole Rivers near their forks, and two over the main Fire Hole Rivers in the Upper Geyser basin. While none of these bridges are very large or costly, all are necessary and serviceable.

Roads.—One road was constructed from near the bridges of the Gardiner, through the East Fork Cañon, *via* the Dry Cañon and forks of the Yellowstone, to Tower Falls—distance, 20 miles.

A road from the forks of the Fire Hole River *via* the East Fork, Mary's Lake, and Mud Geyser, to the foot of the Yellowstone Lake, 30 miles.

Branch of the latter road from Sage Creek by Sulphur Mountain to the mouth of Alum Creek, 4 miles.

Miles.

Aggregate of roads constructed..54

Bridle-paths opened as follows:

Miles.

Paint Pot, length... 11
Passamaria... 22
Painted Cliffs.. 3
Hoodoo or Goblin Land.. 29

 Aggregate of bridle paths constructed............................... 65

Trails constructed:

Miles.

Terrace Mountain... 7
East Gardiner Falls.. 1
Monument Geyser Basin.. 1

 Aggregate of trails constructed 9

The ladders and benches at the Crystal Falls and Grotto Pool, as well as the pole railings to the various points of observation around the different falls, although rude, are convenient and safe for the use of visitors, until a supply of lumber will allow of the construction of better ones. These improvements have been made in addition to the constant care and labor requisite for the removal of falling timber, repairs of bridges, grades, and causeways, and important additions to the latter, notably at Terrace Mountain, Obsidian Cliffs, and Cañon Creek, and a ceaseless vigilance in the prevention of needless forest fires, and wanton vandalism of natural curiosities.

It is believed that the discoveries of the weapons, utensils, and implements, as well as the stone-heap driveways for game, of the present race of Indians or of some unknown prior occupants of these regions, as herein

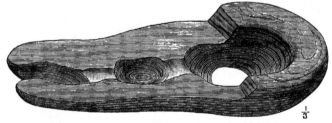

FIG. 29.

illustrated, possess peculiar interest, as well as encouragement for further research; and this is equally true regarding the records, narratives, and traces of early white men in the Park, herein referred to. Nor can it be doubted that the permanent exhibition in the National Museum in Washington of the beautiful pulsating Geyser Cone, from a secluded gorge, and a large collection of geodes, concretions, amethysts, and fragments of fossil timber, obsidian, and other natural objects of interest from various portions of the Park, now in the National Museum, will

 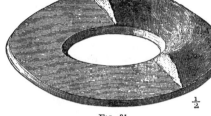

FIG. 30.   FIG. 31.

there greatly assist in disseminating a knowledge, an appreciation, and a desire to visit the enchanting scenery and matchless marvels of the distant Wonder Land. Figs. 29, 30, 31 exhibit curiously-formed water-worn concretions from the Yellowstone Lake, as described in my report of 1880, pp. 16, 17.

IMPROVEMENTS CONSIDERED IMPORTANT TO BE MADE DURING THE COMING SEASON.

Bridges.—As heretofore mentioned, it will be necessary to bridge the Yellowstone twice in order to avoid constructing several smaller bridges over branches, and heavy expensive grades in reaching the Great Falls from Alum Creek. A bridge between the mouth of this stream and that of Tower Creek nearly opposite, at a point where the river is fully 300 feet wide and very deep, but has a sluggish current, gravelly bottom, and fine approaches upon both sides, and another at the narrowest place upon the Yellowstone River below the lake, which is something less than 70 feet between the rocky abutments just above the Upper Falls where there are good approaches, if a bridge be built high above the dashing waters near the brink. A bridge at this point would render accessible far the most open, elevated, and commanding views of the falls and adjacent rapids, as well as the most desirable site for a hotel, application for a leasehold of which by desirable parties is now pending. Several bridges of considerable magnitude, and a number of heavy grades will be necessary in the construction of a road from the Great Falls to those of Tower Creek, where one very high and costly bridge or expensive rock excavation, and probably both, are unavoidable to reach the forks of the Yellowstone, and complete the circuit of roads to the leading wonders of the Park. For reasons heretofore shown, it is very important that the old miner's bridge at the forks of the Yellowstone should be legalized as a toll-bridge, purchased, or else a new one constructed where there are more favorable approaches, as well as another over the East Fork of the Yellowstone near the mouth of the Soda Butte, at that end of the Park, and very long, heavy, and expensive grades or bridges, or both, on the Madison Plateau or Cañon route at the other.

Although not indispensable it is very desirable to construct bridges over the Fire Hole Rivers near their forks, and upon the main fork in, above, and below the Upper Geyser basin, and also just above the midway Geysers as soon as the necessary lumber can be obtained from a mill within the Park. A road from the Excelsior Geyser via the Twin Buttes to the Queen's Laundry, and thence to the forks of the Fire Holes with a bridle-path branch to the Fairy Falls, will be very valuable for its cost, as allowing tourists a choice of routes or a circuitous one upon each side of the river in a trip to the Upper Geyser basin. The desirability of the middle Gardiner Cañon route and of a bridle path to connect with the Two Ocean route to Wind River, the construction of troughs and scaffolding to carry the terrace building waters from the Devil's Thumb to the Liberty Cap for its preservation, and the necessity of a supply of cold water from the McCartney Creek or the West Gardiner in wooden troughs or iron pipes, have been heretofore treated of. Two other matters are of practical importance:

First. The cutting down of at least the dry timber along the main roads and bridle paths to a width sufficient to prevent the annoying obstructions constantly occurring along them.

Second. The removal of the uniformly low but troublesome stumps along the wagon roads, the necessity for both of which will, I am confident, be endorsed by all who have been jolted, or delayed by them. Nor can I believe that the prominent personages who have visited the Park, will consider my views as above expressed in reference to the necessity of additional legislation, registered guides, and an ample police force, far fetched, unnecessary, or impracticable.

SUGGESTIONS REGARDING LEASEHOLDS IN THE PARK.

The clause in the act setting apart the Yellowstone National Park, which refers to revenues from leaseholds for hotel sites and from other sources therein, to be expended in its improvement, renders it evident that it was not the purpose of Congress in dedicating this heritage of wonders as a matchless health and pleasure resort for the enjoyment of our people, to thereby legalize a perpetual drain upon their treasury, a cardinal feature which in the entire management of the Park has been neither overlooked nor forgotten.

But it is also evident that leaseholds cannot be effected to parties possessing the requisite capital and ability to construct and properly manage hotels, which should be adequate to the wants of the public and creditable to the Park, until permanently clear of Indians, and the construction of roads alike necessary for the convenience of visitors, and for the conveyance of a portable steam saw-mill to the proper localities for the manufacture of material for bridge and building purposes.

Hence the undeviating policy has been to encourage and assist in making treaties with the four Indian tribes owning or frequenting any portion of the Park, to cede and forever abandon it as well as the adjacent regions, and with the construction of only such buildings as were absolutely necessary for the safety and convenience of the government officers, employés, and property, crowding the exploration of routes, and the construction of roads, bridle-paths or trails to the leading points of interest throughout the Park; meanwhile making only temporary leases for hotel purposes, but carefully selecting sites and securing propositions for permanent ones.

Upon the accompanying map of the Park may be found in distinct colors the various Fire Hole regions, at which or at other leading points of interest differently colored, the sites properly marked and numbered, as selected for 10 hotels, 2 sanitariums, and 1 for a steamboat harbor and landing at the foot of the Yellowstone Lake, being No. 6 of these hotel sites.

Temporary leases have been made for sites of the hotels at the Forks of the Fire Holes and at the Mammoth Hot Springs, for which as well as for 3 additional sites for hotels, for both of the sanitariums, and for the steamboat wharf, written propositions for permanent leaseholds are now pending, as well as for the establishment of a portable steam saw-mill and zoological garden.

The settled policy of the department has been to grant no titles to any portion of the soil, nor licenses to persons or companies for toll roads or bridges, but rather to make and manage all the improvements of a general nature, such as roads, bridges, bridle-paths and trails, leaving to private enterprise those of a local or private nature, such as hotels, &c., upon leaseholds, under proper restrictions as to time (which, for the purpose of securing a better class of structures, I suggest should be for any period not exceeding 30 years), for a prescribed portion of the frontage for buildings and rear extension for pasturage and fuel purposes at each of these selected sites, leaving the remainder for public use or future leaseholds.

The portable steam saw-mill, together with a sticker planer and other attachments necessary for the proper manufacture of lumber and shingles, should be constructed and managed by private enterprise, under a judicious arrangement as to price, and option of the government as to the place, time, and quantity desired for buildings, bridges, &c., allowing a generous stumpage to the owners of the mill upon any additional quantity which they may wish to manufacture for their own use or for sale to others for the purpose of constructing hotels or other necessary improvements within the Park.

An examination of the accompanying map of the Park, showing the lines of our various roads, bridle-paths and trails, and relative distances, and perusal of the above statements regarding them, it is believed will show a gratifying progress towards the completion of a circle of roads, and a net work of bridle-paths and trails to the main and the minor routes of ingress as well as points of interest throughout the Park, and afford the assurance that appropriations for these purposes need not be perpetual, but that a point is nearly reached when, as above shown, responsible parties will secure leaseholds and make improvements which, without producing great immediate revenues, will soon add to the attractions and enjoyments of the Park, and ultimately at least assist materially in rendering it self-sustaining.

REMARKS ON THE MAP OF THE PARK.

The accompanying map, containing as it does the latest explorations and improvements, is believed to be far the most complete and accurate which has been made of the Park, and will be found reliable in all essential particulars. But as it is intended for practical use in the Park, it is upon a scale so small as to preclude showing many cliffs, cañons, and even some mountains throughout the Park, while the Two Ocean Pass, being outside its limits, is not shown, and the terrible cliffs and yawning cañons beyond the Sierra Shoshone range are mainly omitted in order to show the route of exploration along various creeks in that region. With care it is believed the route of this year's explorations can be traced along a fine continuous line, where, apart from roads or bridle-paths, and save No. 10 at the Two Ocean Pass, each of the 23 camps can be found by their numbers and guidons marked upon the map.

CONCLUSION.

In conclusion, I feel that I cannot in justice fail to express my thanks for the uniform kindness and assistance which I have ever received from yourself as well as from the other officers of the department over which you so ably preside, and it is hoped that any defects in the arrangement or the language of this report may be attributed to the fact that the writer thereof is more experienced in handling the weapons and the utensils of border warfare and life than the pen; but an earnest effort has, by a fair and full statement of facts, been made to show to Congress and the people of the United States, that the slender appropriations which have been made for the protection and improvement of the distant nearly unknown Wonder Land have not been misappropriated or misspent.

My own personal assistants in the Park know full well how thoroughly I appreciate their faithful and earnest services, and need no further recognition than that already made in different portions of this report. Without their cheerful and constant co-operation, my task in exploring and improving the Park, would have been indeed a hard one, and well-nigh impossible.

Very respectfully, yours,

P. W. NORRIS,
Superintendent of the Yellowstone National Park.

APPENDIX A.

ACT OF DEDICATION.

AN ACT to set apart a certain tract of land lying near the headwaters of the Yellowstone River as a public park.

Be it enacted by the Senate and House of Representatives of the United States of America in Congress assembled, That the tract of land in the Territories of Montana and Wyoming lying near the headwaters of the Yellowstone River, and described as follows, to wit: commencing at the junction of Gardiner's River with the Yellowstone River and running east to the meridian passing ten miles to the eastward of the most eastern point of Yellowstone Lake; thence south along the said meridian to the parallel of latitude passing ten miles south of the most southern point of Yellowstone Lake; thence west along said parallel to the meridian passing fifteen miles west of the most western point of Madison Lake; thence north along said meridian to the latitude of the junction of the Yellowstone and Gardiner's Rivers; thence east to the place of beginning, is hereby reserved and withdrawn from settlement, occupancy, or sale under the laws of the United States, and dedicated and set apart as a public park or pleasure ground for the benefit and enjoyment of the people; and all persons who shall locate, settle upon, or occupy the same or any part thereof, except as hereinafter provided, shall be considered trespassers and removed therefrom.

SEC. 2. That said public park shall be under the exclusive control of the Secretary of the Interior, whose duty it shall be, as soon as practicable, to make and publish such rules and regulations as he may deem necessary or proper for the care and management of the same. Such regulations shall provide for the preservation from injury or spoliation of all timber, mineral deposits, natural curiosities, or wonders within said park, and their retention in their natural condition.

The Secretary may, in his discretion, grant leases for building purposes, for terms not exceeding ten years, of small parcels of ground, at such places in said park as shall require the erection of buildings for the accommodation of visitors; all of the proceeds of said leases, and all other revenues that may be derived from any source connected with said park, to be expended under his direction in the management of the same and the construction of roads and bridle-paths therein. He shall provide against the wanton destruction of the fish and game found within said park and against their capture or destruction for the purpose of merchandise or profit. He shall also cause all

persons trespassing upon the same after the passage of this act to be removed therefrom, and generally shall be authorized to take all such measures as shall be necessary or proper to fully carry out the objects and purposes of this act.
Approved March 1, 1872.

APPENDIX B.

RULES AND REGULATIONS OF THE YELLOWSTONE NATIONAL PARK.

DEPARTMENT OF THE INTERIOR,
Washington, D. C., May 4, 1881.

1. The cutting or spoliation of timber within the Park is strictly forbidden by law. Also the removing of mineral deposits, natural curiosities or wonders, or the displacement of the same from their natural condition.

2. Permission to use the necessary timber for purposes of fuel and such temporary buildings as may be required for shelter and like uses, and for the collection of such specimens of natural curiosities as can be removed without injury to the natural features or beauty of the grounds, must be obtained from the Superintendent; and must be subject at all times to his supervision and control.

3. Fires shall only be kindled when actually necessary, and shall be immediately extinguished when no longer required. Under no circumstances must they be left burning when the place where they have been kindled shall be vacated by the party requiring their use.

4. Hunting, trapping, and fishing, except for purposes of procuring food for visitors or actual residents, are prohibited by law; and no sales of game or fish taken inside the Park shall be made for purposes of profit within its boundaries or elsewhere.

5. No person will be permitted to reside permanently within the Park without permission from the D-partment of the Interior; and any person residing therein, except under lease, as provided in section 2475 of the Revised Statutes, shall vacate the premises within thirty days after being notified in writing so to do by the person in charge; notice to be served upon him in person or left at his place of residence.

6. *The sale of intoxicating liquors is strictly prohibited.*

7. All persons trespassing within the domain of said Park, or violating any of the foregoing rules, will be summarily removed therefrom by the Superintendent and his authorized employés, who are, by direction of the Secretary of the Interior, specially designated to carry into effect all necessary regulations for the protection and preservation of the Park, as required by the statute; which expressly provides that the same "shall be under the exclusive control of the Secretary of the Interior, whose duty it shall be to make and publish such rules and regulations as he shall deem necessary or proper;" and who, "generally, shall be authorized to take all such measures as shall be necessary or proper to fully carry out the object and purposes of this act."

Resistance to the authority of the Superintendent, or repetition of any offense against the foregoing regulations, shall subject the outfits of such offenders and all prohibited articles to seizure, at the discretion of the Superintendent or his assistant in charge.

P. W. NORRIS,
Superintendent.

Approved:
S. J. KIRKWOOD,
Secretary.

INDEX.

6 Y P

O

CPSIA information can be obtained
at www.ICGtesting.com
Printed in the USA
LVHW030356091118
596536LV00005B/137